U0069312

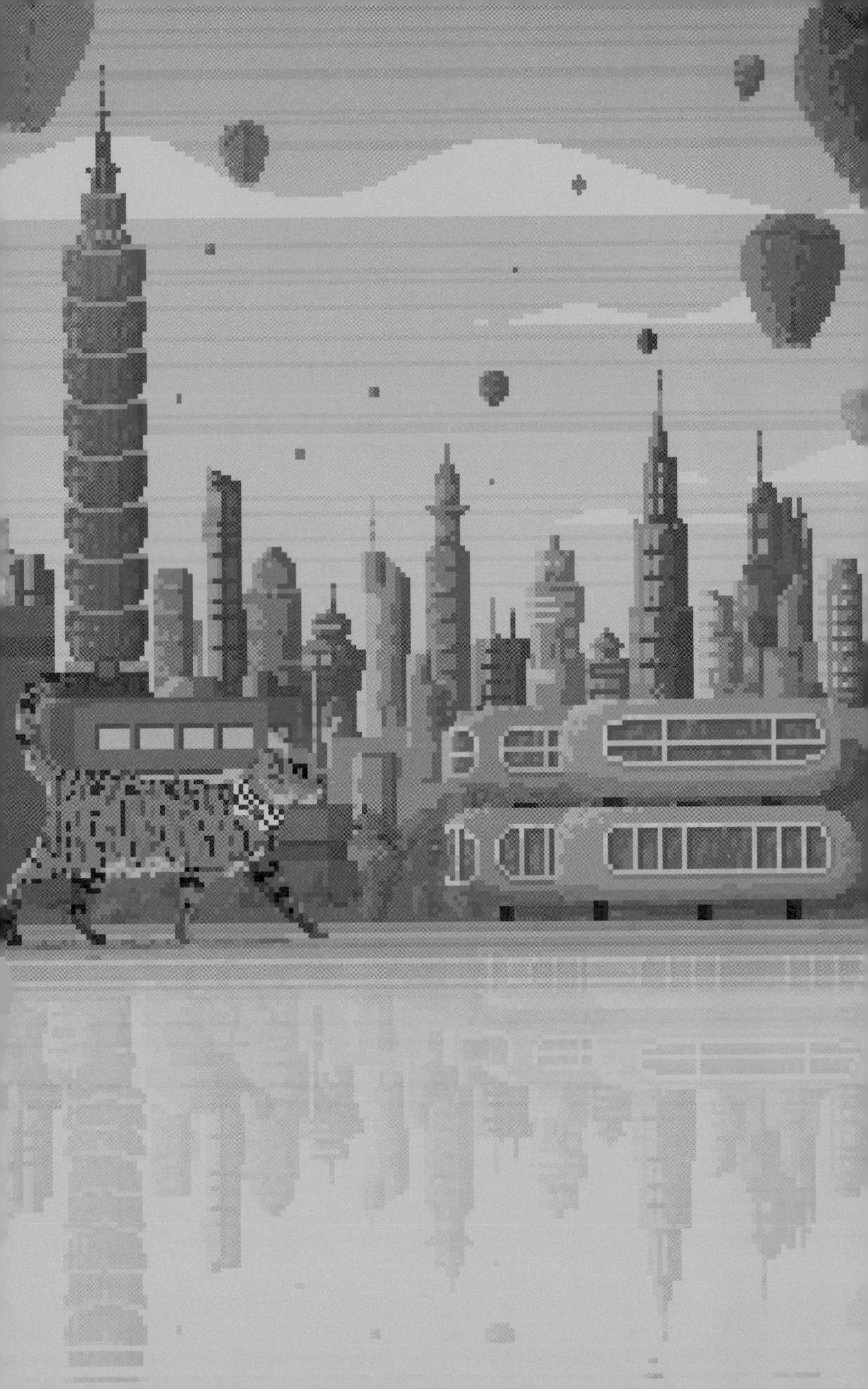

Startup Taiwan:

Foreigners Business Guide

Paolo Joseph L. Lising

Published by Harvest Publishing Co. Ltd

For Mom

About the Book Cover:

The book cover is a playful pixel art rendition that merges the retro and forward-looking vibe of Taiwan from a perspective of a foreigner. It showcases the tension between an old-school mindset, important traditions and the country's ambitions to be even more globally-competitive.

The artwork is by French illustrator, Benjamin Chaumeny also known as ben.ji.art by his thousands of followers on social platforms. Ben.ji has worked on a number of animation films, having earned his Master's Degree from the MoPa - Computer Graphics Animation School.

Contents

INTRODUCTION

What is this book about?

Startup Taiwan: Foreigners Business Guide is an attempt to offer a balanced yet comprehensive guide on setting up a business in Taiwan, with a focus on only the crucial pieces of information that foreign entrepreneurs need.

As Taiwan gains the attention of the international community because of its consistent top-ranked standard of living, landmark policies, and more recently its impeccable model of handling the coronavirus, more foreigners are seeing Taiwan as a place for opening up new businesses.

To be sure, the government has intensified its efforts to lure top-tier foreign talents in a bid to shore up Taiwan's economy by making it more innovative and knowledge-based. Improved regulations are in place and regularly revisited - covering visas, employment, residency, health insurance, taxes, pensions, and many other subject under the Act for the Recruitment and Employment of Foreign Professionals.

To avoid the same pitfalls faced by some guidebooks about founding a business in Taiwan, Startup Taiwan makes sure that it covers not only views from the government, but also exclusive interviews with key resource persons from the private sector and foreign startup founders on their experiences and challenges while establishing their business here.

We covered more than 21 interviews that were included in section case studies, conducted informal surveys of various online foreign groups, and consulted with existing papers and websites on setting up a business in Taiwan – all to offer you a fuller picture of what is going on in the startup scene here.

This book is for you if:

You have an entrepreneurial spirit with a desire to start a business as a foreigner living in Taiwan because you thought you'd just be here for a vacation and now this country interests you as one of the safest countries to live in for foreigners and the right environment for your dream of starting a business.

You are holding a resident certificate and want to continue to live here legally because you no longer want to teach English or do English copywriting jobs for Taiwanese companies a.k.a. "English marketing" anymore, even though admittedly such a job is decent and benefits the future of Taiwan.

You are fresh out of a Taiwan business school and instead of applying for a job here or back home, you prefer to start your own business in this country.

You are having a hard time finding a job because you are from a non-preferred English-speaking country (for teaching English or English copywriting) but feel you can thrive by offering your expertise either with real marketing or some other technical background that you acquired from your home country.

You are considering Taiwan for your plans to live outside your current home country because Taiwan has been in the news lately as a model country and with high standards of living, perfect for your plans to set up your dream business.

You are a policy maker. After learning that this book has some juicy stories from startup founders, you want to read them and see how you could streamline the current policies and procedures.

What's in it for the author?

There really is no money for writing a book, unless your book contains Harry Potter and the like. Even with the previous book that I co-authored, Cyberpreneur Philippines, a finalist for national book awards (clearing my throat), I didn't make much.

This book is inspired only with sincere intentions to help fellow foreigners here in Taiwan, and help promote an underrated country in the international stage. The unbiased approach in writing this book is from my decade of experience as a business journalist with earned recognition (clearing my throat once more).

I can relate to this book. I came to Taiwan to pursue an MBA degree at the National Taiwan University (NTU) back in 2009, found myself in love with this country, and struggled to stay legally. I knew I wanted to start my own business, having one back home with my antique shop, but I didn't know where to start. I observed that while government policies such as in the obtaining of information have improved over time, one key element is still lacking: the effective communication of these new policies, for optimal benefit to the economy.

What's in it for me? My bigger purpose is to provide an educational platform for entrepreneurs from developing countries through my startup, MillionDC.

This book, for example, is educational material for entrepreneurs seeking to flourish in Taiwan. At MillionDC.com, we hope to provide more tailored educational materials for countries such as the Philippines, my home country, and Indonesia which is close to my heart. Meanwhile, apart from the guides in starting a business in Taiwan, I have also dedicated two chapters about our experiences while building MillionDC and I think these would be useful for a founder who is solidifying his business concepts.

Use #StartupTaiwan

If you find this book useful, please promote it to your friends and network on social media using the hashtag #StartupTaiwan. This is our digital campaign that will direct the attention to the bigger goal of promoting Taiwan to foreign talents, even to those global Taiwanese who are considering coming back home.

Paolo Joseph L. Lising,

Author

CHAPTER 1

Why Start a Business in Taiwan?

Chapter Snippets:

This chapter looks into the first touchpoints for considering starting a business in Taiwan. It first covers various international surveys detailing business-related ranking, followed by the topline policies in allowing foreigners to set up enterprises in the country. Lastly, we look into how the national government plans to remain competitive worldwide through enhancing the country's startup ecosystem.

Reasons to start a business in Taiwan

Doing business in Taiwan has become much easier with the country's excellent living conditions and the adoption of economic deregulation policies in recent decades. As such, foreign investors can take advantage of transparent laws on foreign investments. These, in turn, make the business landscape friendly for interested parties. The perks? Well, here are some of them:

There are no restrictions on foreign ownership. Thus, you are free to set up your business in this wealthy nation. Also known to have the 10th highest economic freedom score (77.3 according to the Economic Freedom of the World Index as of 2019), Taiwan boasts of its free-trade policies, allowing investors to choose whom they will trade with, cooperate with, and compete with.

Your intellectual property rights (IPR) are well protected. Doing business in Taiwan allows you to enjoy fair play in terms of IPR, given that the country invests heavily in science and technology. The World Bank has included it in the list of top countries for ease of doing business.

Taiwan is where expats feel safe. In fact, according to InterNations Expat Insider 2019, the country is ranked as the best place to live in for expats out of 64 destinations, due the excellent quality of life, affordability of healthcare, transportation infrastructure, and other similar factors.

"Since first featuring in the survey in 2016, Taiwan has consistently ranked in the top 5. In 2019, it manages to reclaim first place.

Taiwan offers a great quality of life (3rd out of 64 destinations), favorable personal finances (6th), an impressive working life (8th), and good results for ease of settling in (14th)," InterNations Expat Insider 2019.

11th World Competitive Country. Taiwan has been ranked 11th-most competitive economy in the world and third-most in Asia, behind Singapore and Hong Kong in the 2020 annual world competitiveness rankings released by the International Institute for Management Development (IMD).

Topline government policies in starting a business in Taiwan

Any foreign entity with a business agent or a fixed place of business in Taiwan must register as a legal entity in this country. Moreover, registering a business in Taiwan is pretty much straight-forward but can take two to three months. The step-by-step ways to register is on the next chapter.

Consumer protection and fair trading

To protect consumers' interests, the Taiwanese government legislated the Consumer Protection Law in 1994, together with the creation of the Consumer Protection Commission to serve as an independent regulator under the Executive Yuan, the executive branch of the government of Taiwan. This commission comes up with regulations to educate consumers on their rights, handle consumer complaints, and protect them from false advertisements, Internet scams, and other fraudulent issues. Predominantly, there are no price regulations in Taiwan. However, public transportation and utility prices are regulated and closely monitored by the government.

Import and export regulations

To transport any items to or from Taiwan, you must have an import/export permit. There are stricter rules for importing and exporting specific strategic technologies and you need to refer to the "Regulations Governing Expert and Import of Strategic High-Tech Commodities." If you are looking into importing controlled commodities, you first need to secure an International Import Certificate with the Bureau of Foreign Trade (BOFT). The BOFT is also in charge of approving permits for exporting strategic high-tech commodities.

Land use

According to Article 18 of the Land Law, a foreigner or business may acquire land in Taiwan provided that the foreigner's home country also allows the same right to Taiwanese citizens.

Foreigners are permitted to acquire land for the following purposes:

i. Investments (usually for animal husbandry and agricultural industries, major infrastructure projects, and overall economic development)

ii. Personal use (residences, business offices, shops, factories, etc.)

iii. Public welfare purposes (cemeteries, hospitals, churches, diplomatic and consular buildings, schools for children of foreigners, etc.)

Meanwhile, according to Article 17 of the Land Law, foreigners are prohibited from leasing, encumbering, or being transferred lands for agriculture, pasture, fishery, or hunting; water sources; salt or mineral deposits; military areas and lands adjacent to the national frontier.

Foreign exchange controls

The government of Taiwan balances the outward and inward remittance of funds. A resident may remit up to USD 5 million outside Taiwan per year without approval from the central bank. Banks in Taiwan are required to seek consent from the central bank prior to any single remittance beyond USD 1 million. A resident corporation may remit up to USD 50 million outside Taiwan per year without obtaining approval from the central bank.

Any overseas investment beyond USD 50 million requires approval from the Investment Commission of the Ministry of Economic Affairs (ICMOEA). Likewise, any investment in Taiwan, of whatever monetary value, must have prior ICMOEA approval.

Tax incentives

There are many tax-related incentives under the Act for Development of Small and Medium Enterprises, the Statute for Industrial Innovation (SII), and the Act for the Development of Biotech and New Pharmaceuticals Industry. Below are the tax incentives in general:

i. Tax credits for Research and Development (R&D)

ii. Special tax deferral for incurring capital for contributing technical know-how

iii. Special tax incentives for hiring Taiwanese staff or raising the salary of local staff.

A single annual application R&D Investment Tax Credit should be made within four months prior to the corporate income tax return filing due date. A foreign company may elect any of the following methods to calculate R&D credits.

Companies can get up to 15% tax credit through R&D tax credit for qualified R&D expenditures against its corporate income tax for the current year. The tax credit must, however, not exceed 30% of the business income tax for that year and may not be carried over to succeeding taxable periods.

Companies may get a tax credit of up to 10% of acceptable R&D expenditures from its corporate income tax for the current year, with unutilized R&D tax credits carried over two consecutive years once the 30% limit on the corporate income tax has been exceeded.

Non-tax incentives

Through non-tax incentives, newly built companies get to lessen their operating costs. Such incentives include the following:

i. Land lease incentives in industrial parks

ii. The Industrial Technology Development Program

iii. Low-interest loans

iv. Government participation in investment

Foreigners are generally not restricted from investing in the various industries in Taiwan, except when these involve funds coming from mainland China or when national security concerns require participation to be limited. There may be some instances when laws would limit the percentage of equity that companies and foreign nationals can possess in specific industries (such as telecommunications, posts, and shipping) to adhere to decrees in line with national interests in the social, economic, or cultural arenas.

Action plan for enhancing the country's startup ecosystem

The National Development Council (NDC), the policy-planning agency of the Executive Yuan (EY) formed in 2014 by streamlining and merging six previous government agencies, has a mandate to accelerate Taiwan's innovation and entrepreneurship, with the goal of spurring economic growth.

In 2018, the EY approved NDC's Action Plan for Enhancing Taiwan's Startup Ecosystem. "By pushing ahead [with] this Action Plan, the Government would create a robust startup ecosystem in Taiwan, raising Taiwan's visibility internationally, and making Taiwan Asia's startup nation," NDC said.

To be sure, Taiwan's real GDP growth rate has not hit double digits since a record high of 10% in 2010. In fact, the growth rates in the following years were less than 5%, and only around 2% in the last four years.

NDC's action plan to help spur Taiwan's economic growth is comprised of five strategies, namely: providing ample early-stage funding for startups; developing talent and adjusting regulations; building partnerships between startups and the government; providing startups with various exit channels; and helping startups tap into global markets. These strategies are to be jointly implemented by 13 government agencies.

The action plan includes giving tax incentives for angel investors, and cooperating with international venture capital firms (VCs) to invest in "forward-looking" industries such as artificial intelligence (AI), internet of things (IoT), augmented or virtual reality (AR/VR), and the biomedical industry.

It also includes implementation of the Startup Regulatory Adjustment Platform, a single window that would address concerns of startups in a proactive approach of studying and including international best practices in existing laws that cover the startup industry. It also implements the government open data platform where startups can freely search for any government data, and use them for their research.

Augmenting talent base with foreign recruits

One of these action plans is for attracting and retaining foreign talents - the implementation of the Act for the Recruitment and Employment of Foreign Professionals. This Act, approved in 2017, aims at raising national competitiveness by enhancing the recruitment and employment of foreign professionals. This Act seeks the relaxation of regulations on work, visas, and residence; easing of provisions concerning stay or residence of parents, spouses, and children; and providing retirement, insurance and tax benefits.

Meanwhile, the government is also encouraging global Taiwanese to come back home and invest in Taiwan with its "Action Plan for Welcoming Overseas Taiwanese Businesses to Return to Invest in Taiwan."

NDC said the plan is oriented for the needs of homecoming enterprises, actively assisting Taiwanese companies in returning to Taiwan by providing customized single-window services for land, water and electricity, manpower, taxation and capital.

Plans to create Asian Silicon Valley

An earlier plan by NDC was Asia Silicon Valley Development Plan that was launched in September 2016 aimed at "connecting Taiwan to global tech clusters and create new industries for the next generation."

The plan seeks to promote innovation and R&D for devices and applications of the internet of things (IoT), and improve Taiwan's startup and entrepreneurship ecosystem, according to NDC website.

The plan has four implementation strategies namely; optimizing Taiwan's startup and entrepreneurship ecosystem by increasing talent supply, providing business expansion capital, and adjusting laws and regulations; enhancing linkages with renowned clusters worldwide by forging connections with the R&D capabilities of Silicon Valley and other global innovation clusters; building complete IoT supply chain by integrating Taiwan's hardware advantages into software applications; and building sites for smart products and services including establishing a quality internet environment.

The plan's implementation period will run through 2023 and a budget of NTD 11.3 billion (USD 350.4 million) has already been allocated. The Asia Silicon Valley plan combined with other digital economy plans of the government will hopefully raise Taiwan's IoT global market share from 3.8% in 2015, to 4.2% in 2020, and to 5% in 2025. A story by Taipei Times said the plan was not formed to make a Taiwanese copy of Silicon Valley in California, "but to encourage Taiwanese industries to have the spirit of innovation.

Plans to improve English

It seems that the government has been reading through various discussions on social platform groups of foreigners in Taiwan, the NDC has formulated a blueprint for developing Taiwan into a bilingual nation by 2030.

NDC said the blueprint would be "elevating national competitiveness" and "cultivating people's English proficiency".

Some of the ways to achieve this goal according to NDC include promoting English TV channels, and encouraging the Taiwanese channels to produce and broadcast English programs. Related to founding a business, NDC suggested to have government documents in English, improving the English language capabilities at the science and industrial parks, and promoting government websites with English versions, among others.

The NDC noted that the current blueprint is different from previous bilingual policies. For example, it is designed to enhance the nation's overall competitiveness in using the language rather than simply the ability of the students to pass examinations. The blueprint also has a focus on enhancing people's English proficiency and encouraging not only students but the entire nation to learn the language.

CHAPTER 2

Doing Business in Taipei

Chapter Snippets:

Taipei - Taiwan's capital and hub of business, finance, and technology - driven by foreign direct investments since the 1960s that brought the so-called Taiwan Miracle, deserves a separate chapter in this book. We begin by introducing why foreigners decide to settle in this city, followed by a look at the available Taipei government support for foreign startups, and a section on the city government plans to stay globally competitive under its Design City; Dream City campaign involving thinkers from around the world. Lastly, we included an interview with the Invest Taipei Office that serves as the first window to any foreigners planning to set up their business in the city capital.

Why Taipei?

Numbers don't lie. Taipei is ranked number one as the best city to live in as an expat, according to a survey of 20,000 expats living in 187 countries by InterNations, a Munich-based social network for expats around the world. Taipei scored particularly high in overall quality of life according to InterNation Expat City Ranking 2019 that looked into the following indices: quality of urban living, getting settled, urban work life, and finance and housing.

To be sure, 98% of those who were surveyed said they were satisfied with the local transportation. In a separate citation, Insider Inc. published an article on Taipei Metro noting that "*... Taiwan, metro is cheap, clean, and efficient... The rest of the world could learn from Taiwan's subway.*"

InterNations' survey also noted that 94% of the respondents were happy with the availability of healthcare in the city. Taipei's healthcare is tied with the Taiwan National Healthcare Insurance. Singapore-based bio and health science publication, BioSpectrum, has listed Taiwan in top ten worldwide for healthcare. "Taiwan's healthcare model imbibes 6 main strengths; high quality, affordability, technology and innovation, patient-oriented services, comprehensive specialties and services of leading physicians," BioSpectrum noted in its December 2019 article.

InterNations' survey showed that 96% of the respondents had overall feelings of safety while 80% found the local residents generally friendly. A USA Today report listed Taipei in the top ranks of safest cities, citing its health and personal security. For example, life expectancy at birth in Taipei is 83.6 years, more than 10 years longer than the global average.

For foreigners setting up their businesses, it is also worth noting that Taipei ranks 6th in a global ranking of cities based on inclusive prosperity. The 2019 Prosperity and Inclusive City Seal and Awards (PICSA), commissioned by the Basque Government and compiled by Australian-based consulting firm D&L Partners, cited Taipei's rich, export-driven economy and a high-tech sector that leads the world in the communications and information technology industries.

Taipei City's support for foreign businesses

The city government of Taipei has established the 'Taipei Municipal Self-Government Ordinance for Industrial Development' as its vehicle to support the establishment of businesses in the city.

Foreigners may apply under Taipei's Subsidies & Incentives for Taipei Industry for those business that require research and development (R&D) support. "In order to encourage technology development, Taipei City is the first city government who provides R&D subsidies which can be up to 5 million NT dollars," according to Taipei City's official website.

The city government also provides up to NTD 5 million for brand establishment; up to NTD 3 million for incubation; and up to NTD 1 million for new startups.

Here is a complete checklist to complete your application for Taipei government subsidy.

i. Complete application form

ii. Business plan for new investments

iii. Business Issuance Registration Certificate or Capital Increase Certificate

iv. Uniform invoice or other documentary proof of new equipment purchase or technology transfer

v. Financial statements for the three most recent years certified by a certified public accountant. For companies established less than three years, all certified financial statements since the establishment shall be provided; for small and medium enterprises, financial statements compiled based on source documents approved by tax collection authorities may be provided as an alternative

vi. Business profit tax return and Annual Income Tax Return Forms for the most recent year. For companies established less than one year, requirement for submitting Annual Income Tax Return Forms may be waived

vii. Documents demonstrating applicant does not have overdue taxes and fines issued by the National Taxation Bureau of the Republic of China (Taiwan)

viii. Documentary proof of authenticity of items submitted hereunder for application of subsidies

ix. Other documents required by DOED

FAQs on Taipei subsidies

The government of Taipei, in many of its websites, has enlisted a number of frequently asked questions (FAQs) on doing business and we have picked and reworded some of those questions and answers that are relevant for foreigners who are planning to set up a business in this city.

Does the Taipei City government provide free accounting and legal consultation services?

YES. Aside from the Startup Clinic Service provided by StartUP@Taipei Office, the Taipei City government also offers on-site accounting and legal consulting services for free. You may visit its Accounting Service Center on the first floor of Taipei City Hall. It's on the North Wing, Office of Commerce Counter No. 6. Call them at telephone number 02-27596019.

Are foreign investors qualified to apply for incentives and subsidies under the Taipei Municipal Self-Government Ordinance?

YES. Foreign investors can be qualified applicants, if they get the investment permission from the Ministry of Economic Affairs in accordance with the Foreign Investment Regulation. The foreign investors can be sole-proprietor or partnership companies registered in Taipei City. There are no specific requirements on the applicant's industrial classification.

What types of Taipei government subsidies can a Branch of Foreign Company for?

You may apply for all of the items listed in the Taipei government support for doing business in Taipei section of this book, specifically R&D, investment incentive, branding subsidy, and innovation incubation and angel investment (startups). The subsidy could cover: newly hired employee's payroll; annual housing tax and land-value tax; rental subsidy for providing for the assets located within the city; interest payments; and city-owned real estate, among other items.

What are the qualifications for the startup subsidy?

According to Article 10 Section 2 of Taipei Municipal Self-Government Ordinance for Industrial Development, applicants could apply for the startup subsidy if they meet the following requirements: The company or foreign company must be registered in Taipei City; the startup plan should be innovative, creative or with potential; the startup plan is not subsidized by any other government unit; and the subsidy should not exceed 50% of project expenditure, with NTD 1 million being the largest-possible subsidy.

Could an individual apply for a startup subsidy? Is there any age restriction?

NO. Therefore, please complete your company registration to fulfill the qualification. Moreover, there is no restriction about age.

If all the staff are foreigners, could their salaries be subsidized?

YES. However, please note that only a full-time employee (labor or job insurance proofs need to be provided) who actually participates in the project could be subsidized for his/her salary.

If we want to apply for a startup subsidy of NTD 1 million, should we prepare our own NTD 1 million in advance?

NO. However, at the end of the project, you need to prepare all the receipts or invoices for the total budget, including government- and self-funding.

Is there any restriction on industrial classification for any company applying for innovative research and development subsidies?

NO. To encourage industrial innovation, research and development, any investor who is engaged in technology and new service development can apply for the subsidies to pay for the required expense with no restriction on industrial classification categories.

If we have already applied for the subsidy this year, could we apply for it next year after a capital increase?

YES. As long as your company is a small or medium enterprise registered in Taipei City, or it's a new startup, you are qualified to apply.

If a company has already applied or obtained subsidies from other government entities, can it still apply for incentives or subsidies with the Taipei city government?

NO. If the same or similar project has been granted an incentive or subsidy by another government entity, the investor shall not be allowed to apply, in accordance with "Taipei Municipal Self-Governance Ordinance for Industrial Development".

Who is eligible to apply for the Taipei government's industrial development subsidy? Can a small- or medium-sized enterprise be granted a subsidy under this category? Are companies registered in other cities/counties eligible for this subsidy?

YES. This subsidy is available to companies, sole proprietors or partnerships registered at this city in accordance with the law, and there are no restrictions to the industries. SMEs or companies registered in other cities or counties are not eligible for this subsidy.

Is there any incentive provided by the Taipei City government for the private sector to participate in public projects?

YES. According to "TCG Municipal Guideline for Encouraging Civil Participation in Public Infrastructure Projects through Land Value Tax, Building Tax and Deed Tax Abatement," the private sector can enjoy local tax incentives, such as exemption of land value tax for five years, 50% reduction in building tax, and 30% reduction in deed tax.

Future plans for Taipei

In keeping the city globally competitive and innovative, the city government has launched "Design Taipei; Dream Taipei" as an action plan to turn designs into profitable concepts not just within the art industry but across all enterprises, similar to MillionDC's vision for developing countries.

According to TaipeiIecon.com, the website detailing Taipei's International Design Awards, the city government is hosting competitions in search of "creative designs with great commercial potential...

The competition also serves the purpose of enhancing urban lifestyles through innovative design concepts." Furthermore, its 2019 Taipei International Design Award has the theme "Design for an Adaptive City" to encourage designers to show concern for social issues and contribute to urban prosperity and progress through their designs. Participants may enter into the following categories: design, applied design, universal design, industrial design, visual communication design, and public space design. The broader set of categories is aimed at finding even more promising designers not just from Taiwan but also from all over the world.

Case interview: Where to Start in Taipei

The Taipei City government's Department of Economic Development established the Invest Taipei Office (ITO) as the first touchpoint for any startups, either from another city or from another country, wishing to establish their business in Taipei. To give light on Taipei's overall policies for startups, we interviewed Robert Lo, the Executive Director and Head of Invest Taipei Office, who also holds a position of Director at the Industrial Technology Research Institute (ITRI).

"Treat ITO as your first contact window... any foreigner can walk in our office [for consultation]. We are a problem-solving office, [set up] in order to overcome barriers [in] understanding Taipei City rules," Lo said.

He explained that the approach of ITO "is more reactive." For example, while the national government provides a lot of resources and incentives, these incentives are "very fragmented" and "not very easy [to understand] for those first-time visitors from abroad."

Attracting tech-oriented entrepreneurs

"Our main focus is to recruit talented/qualified tech-oriented foreign entrepreneurs to start business here in Taipei, either to set up an office here or explore any business opportunities here in Taipei... The main priority is software [engineering] talents, which would be aligned with one of our national six strategies: [the] cybersecurity industry," Lo said while also noting that his office has helped almost all types of foreign-owned startups.

One of the ways to establish a startup in Taipei is by accessing the city's core strength in manufacturing as well as its pool of talented software engineers.

"Taiwan has a good reputation [globally] in terms of advanced manufacturing and our respect for intellectual property... We also provide good software talents, considering that the biggest tech companies - Google, Amazon, and Facebook - opened their software R&D center here. These companies even opened up 3,000 vacancies to recruit Taiwanese talents, to create their Asia-Pacific hub, to provide service all over the world," Lo noted.

He said these highly qualified software developers could work not only with big companies but also with smaller foreign startups on a hybrid model where their main office could be located in their home country with an engineering team located in Taipei via a branch office.

When asked about ITO's key performance indicators, Lo said that apart from attracting early-stage startups, Lo said that the city is also trying to attract innovative SMEs and large corporations in hopes of increasing capital investments. "We would like to increase the amount of capital injected [into Taipei] in billions annually. [For example], if they try to set up their subsidiaries or branch office here, they need to inject capital," he said. He said the second indicator would be job creation that could result from entrepreneurs registering their business. The third indicator is the amount of potential versus realized investment including additional investments by companies that were already registered in Taipei.

Talents Taipei

Lo said Taipei's goal to increase foreign talents is carried out partly through the Talents Taipei program, a 10-day business development soft-landing program to explore Taiwan's market potential for high-tech foreign startups.

"It's a mini-accelerator program that Taipei has been running for three years, now on its fourth," Lo said. He said 30 entrepreneurs are selected to be part of this program each year. For the selection, the government forms a jury that includes a professor from the National Taiwan University and from Academia Sinica who would check if the applicant has profound knowledge in his startup field. Lo said the jury also includes an executive from Google and a VC from Taiwan or Silicon Valley to check whether the proposed idea or startup has some application in Taiwan and if the product or service is aligned with market trends. To be sure, Taiwan has already set up its innovation center in 2014 and Lo was the center's first CEO.

Challenges for attracting foreigners

"The regulation, in terms of compliance requirements, is not so friendly for foreign talent [that] would like to do serious business here [in Taiwan]. If you try to open even a small legal entity, there are still a lot of requirements for you to provide a lot of paper works... In the US, you can register your company in one day," Lo noted.

He said that this concern has nothing to do with local government. "It should be [addressed] by the financial commission who supervises the banks, and foreign investment committees to check the capital... Our government should open online registration of legal entities, to automate each step from a website," Lo noted.

Lo also aired his concerns on Taiwan's capital. "For entrepreneurs, it's ridiculous to track the small amount of capital injection," he said, citing that capital injected to Taiwan needs to go through due diligence, requiring the entrepreneur to provide notarized original documents. "I don't think that is necessary," Lo noted.

CHAPTER 3

Basic Business Structure in Taiwan:

Chapter Snippets:

For foreigners who want to register their companies on their own, this is a step-by-step guide, with insights from agencies and foreign business owners who have done it. It breaks down the types of businesses that you can establish in Taiwan and then jumps straight to the order in which you can register and start your company. For foreigners who are outside Taiwan, this guide helps you learn the process while negotiating with an agency who will help you do the registration on your behalf.

Once you have already understood Taiwan's business environment, it's time to put structure to your enterprise and register it. Learn how to hire yourself, your co-founders, and employees. Follow the rules and have a hassle-free setup for your business.

Define the type of business entity you want to set up

According to accounting firm FDI China, the most common types of foreign entity registrations are the Branch and the Subsidiary (Limited Liability Company and Company Limited by Shares). Below are the agency's definitions and insights on each type of entity. Some of the points were reworded and rearranged for clarity. We've also included some insights, where appropriate, from the 'Doing Business in Taiwan' document by business adviser Grant Thornton, Taiwan.

Branch:

A Branch structure is not considered as a legal entity. It operates on behalf of the parent company, meaning that the net profit after income tax payment can be retained to the parent company without additional income taxation. In addition, a Branch entity appoints an individual as its agent for litigious and non-litigious matters, and a branch manager.

FDI China's tip: We recommend to our clients to register a branch when the intention is to operate in regulated industries such as banking and financial services.

Requirements and benefits for setting up a branch

- Company Name: Translate parent company's name for inspection
- No minimum capital requirements
- Permission for business / import and export
- Direct issuance of sales invoice to the client by your branch
- Payment of expenditures and inventory purchasing
- Enterprise income tax: 17%
- Business bank account required
- Physical office required

"While many foreigners choose to invest in Taiwan via branch offices of foreign companies for tax reasons (branches are exempt from withholding tax on repatriated income), anyone seeking to open a branch in Taiwan should take precautions to limit the assets housed in the branch's foreign head office," Grant Thornton, Taiwan.

Subsidiary: Limited Liability Company (LLC)

A Limited Liability Company (LLC) structure is a separate legal entity and independent domestic company. In Taiwan, an LLC has after-tax (17%) operating income, which can be wired out to the parent company by paying 20% withholding tax. In addition, this type of company must have at least one shareholder and director.

FDI China's tip: We recommend to our clients to register a Limited Liability Company (LLC) in Taiwan for foreign investors due to its flexibility and reduced administrative requirements. Multinationals usually set up this type of business structure.

Requirements and benefits for setting up an LLC

- Company Name: need to register a new name in Taiwan
- No minimum capital requirements
- Permission for business / import and export
- Statutory audit / annual tax required
- Enterprise Income tax on distributed earnings: 20%
- Enterprise income tax: 17%
- Business bank account required
- Physical office required

Subsidiary: Company Limited by Shares (CLS)

A Company Limited by Shares structure is a separate legal entity and independent domestic company. Unlike the Limited Liability Company (LLC), this entity is limited by shares; meaning that it is the shareholders' obligation to pay the company for the shares they own. In addition, it needs to be formed by two or more shareholders and at least three directors.

Requirements and benefits for setting up a CLS

- Company Name: need to register a new name in Taiwan
- No minimum capital requirements

- Permission for business / import and export
- Statutory audit / annual tax required
- Enterprise Income tax on distributed earnings: 20%
- Enterprise income tax: 17%
- Business bank account required
- Physical office required

Difference between the CLSs and LLCs

CLS must have at least two shareholders. This type of corporation must have at least three directors, but it does not have a maximum number of shareholders. For LLCs, having one shareholder is sufficient, but the maximum number of shareholders is only 50.

There is no minimum capital requirement for CLS and LLCs unless the entity is involved in an industry with a required minimum capital set by the competent authority or if it is going to employ a foreigner to work in Taiwan.

Representative office vs Sole partner

Meanwhile there are other business entity types in which you can register your company: a representative office and a sole partner.

A representative office is a separate legal entity representing a foreign company in Taiwan. This business solution is known as an easy and inexpensive method to establish a legal entity in the country. This is ideal for foreign businesses which do not intend to conduct actual business in Taiwan but want to source data and acquire goods from the country. This is also suitable for startups and businesses that offer offshore trade or consultancy services.

The Statute for Investment by Foreign Nationals (SIFN) also permits foreigners to invest in Taiwan on their own or through partnerships with one or more people. These partnerships, however, are limited. They are not allowed to hire foreigners as workers and all partners are required to submit a personal income tax return. A limited partnership in Taiwan is considered a tax-transparent entity but it is not subject to corporate income tax. Whether it is a limited company or a company limited by shares, there is no minimum capital requirement for both limited partnerships and sole proprietorships.

6-point step guide to register your company in Taiwan

Step 1 - Name your business

Prepare Chinese and English names for your company. Check with the Department of Commerce, Ministry of Economic Affairs (MOEA) if your intended company name in Chinese is already taken.

If it helps, consult with a Taiwanese friend to check for any cultural tones or sensitivity with regards to your intended Chinese name. The finished product of this step is a document Copy of Preliminary Check of Firm's Intended Name. You have to complete the full registration process within six months to be able to keep your intended name for your company.

Once you get your company name registered, you will need to get your company stamp and your own name stamp as well. These two stamps will be your official signatures for signing any document for your company.

Step 2 - Secure investment approval

Seek an approval of your investment with the MOEA Investment Commission (MOEAIC). You will need to specify the amount that you will be investing and where the money is coming from. For anti-money laundering purposes, you will also need to provide bank records for holding the amount of your intended investment.

It is easier if you're an Alien Resident Certificate (ARC) or Alien Permanent Resident Certificate (APRC) holder and that the money you invest comes from your Taiwan bank account, say from your salary from your full-time work in the country. Your

Taiwan bank can prove that you have that money that you intend to use for your business, and it can show that you have obtained it from your salary.

NOTE: At this point, you should start looking for your official business address. You cannot use your home address as your business address, unless you have this agreement in writing with your landlord. You cannot use a government P.O. Box as your address, such as those offered by Taiwan Post Office. We once made this mistake at MillionDC. use a virtual business address instead, which costs NT3,000 per month on average with 12-month contracts. Choose your company address wisely. If you intend to apply for a Taipei government grant, your business address / virtual address must be registered in Taipei.

Below are the documents that you need to secure investment approval:

- Copy of Preliminary Check of Firm's Intended Name (from Step 1)
- Application form
- Bank statements
- Copy of your passport

Step 3 - Open a bank account (temporary)

This step could be done parallel to Step 2 but be sure you have completed Step 1.

Make sure that your bank account is under the name stated in your *Copy of Preliminary Check of Firm's Intended Name*, and you are stated as the owner of the company.

The bank account that you just opened will remain temporary until finishing Step 4.

NOTE: Do not transfer your investment money to your temporary account until after you have secured investment approval (Step 2). When applying for this temporary bank account, bank officers will simply ask you to deposit NT1,000.

Some banks will ask for your office address. Remind them that this step will only be necessary after getting investment approval (Step 2) and when you secure company registration (Step 6).

Below are the documents that you need to secure investment approval:

- Approval Letter/Document of Your Firm's Intended Name (from Step 1)
- Application form (each bank has its own application)
- Copy of passport and ARC
- Company and personal stamp

Step 4 – Transfer your investment money to your temporary bank account

Only do this step when you have obtained the approval letter from MOEAIC for foreign investment.

IF YOU ARE NOT AN ARC/APRC holder, inform your temporary bank that you will be transferring money from your hometown country as "Direct investments by foreign nationals and overseas Chinese" before proceeding with transferring the exact investment amount that you specified in Step 2.

IF YOU ARE TRANSFERRING MONEY FROM ANOTHER TAIWAN BANK ACCOUNT, it is highly suggested to do it over the counter so that you could get a paper copy of your transfer record / form that you will need for Step 5.

Keep the original transfer or remittance form as you will need it for the final Step 6.

Step 5 – Apply for Capital verification approval

Below are the documents that you need to apply for capital verification approval:

- Capital verification form
- Temporary bank account (cover page, page with updated balance, and the page with manager's seal)
- Original transfer or remittance form (or transfer forms for those who transferred money from another Taiwan bank)

Bring all the above documents to MOEAIC Address: 8F, No. 7, Sec. 1, ZhongZheng District, Roosevelt Rd., Taipei City, Taiwan (R.O.C.), Tel: 886-2-3343-5700

NOTE: At this point, you must have a signed contract for your company office / virtual office lease. You will need this lease contract for turning your temporary bank account into an official bank account for your company.

Step 6 – Apply for company registration

You're almost done! Once you have accomplished Steps 1- 5, you will just need to file a formal registration application for your business. Below are the documents that you need to apply for company registration:

- Approval Letter/Document of Your Firm's Intended Name (from Step 1)
- Application form
- Lease agreement (for your office / virtual office)
- Copy of passport and ARC
- Company and personal stamp

Bring all the above documents to MOEAIC Address: 8F, No. 7, Sec. 1, ZhongZheng District, Roosevelt Rd., Taipei City, Taiwan (R.O.C.), Tel: 886-2-3343-5700

REMINDER: Follow the timeline

Under Article 10 of Taiwan's Companies Act, a company may be dissolved under these circumstances:

- The company fails to start its business operation after six months from its registration, unless it has made an extension registration
- The company has discontinued for more than six months after starting its business operation, at its own will, unless the company has made the business discontinuation registrations

Chapter 4

How to Fund your Startup in Taiwan

Chapter Snippets:

This chapter explores the many ways to fund a startup in Taiwan. We begin by introducing the funding concepts ala "Business Funding 101" followed by a comprehensive list of available funding offered by the Taiwan government both at local and national levels. It is ideal for companies to set aside a good runway of one year to keep your business afloat while applying for government grants and subsidies. For private sector funding, we have also included a comprehensive list of active VCs in Taiwan. We have included interviews with VCs.

Business funding 101

There are four ways to fund your business.

1. Your own money

This is quite self-explanatory. If you have enough money, you can start a legal business in Taiwan right away. The amount of money that you will need will depend on the kind of business that you want to start.

The core advantage of using your own money is the “freedom” to use it without conditions set by another person. No one will blame you if you don’t earn, and when your company gets big, you get all the profit.

The downside, especially for those with limited savings, is that you may have to spend less for your food, house, electricity, and gasoline while your startup company is still starting. You could lose your savings if your company is not successful.

2. Loan from a bank or creditor

Foreigners think that it’s hard to borrow money from banks in Taiwan. However, through the government’s Small Enterprise Loan, foreign companies can avail of up to NT5 million in loans (see item 7 under List of National and Local Taiwan Government Funds for Startups in this chapter). Borrowing from a bank requires a lot of paperwork to prove that you’re able to pay loans.

3. Selling a part of the business that you want to start

You can sell a part of your business to someone who is willing to give you the amount of money that you need and he will become a part owner of your company. Equity is the part or portion or percentage of the company that you will start. It can be 5 percent, 10 percent or 50 percent. Remember that this means you will not be the only owner of your company and a portion of your profit will be taken.

Nc matter how you sell part of your business, you will need a document that details the conditions of equity financing such as a Term Sheet, a Seed Investment Agreement, a Shareholders' Agreement, or a Share Certificate.

4. Crowdfunding

Many companies have successfully started their businesses by using crowdfunding. The simple meaning of crowdfunding is to get your money from the public so that you can start your company.

For example, MillionDC has raised funds to build the very first version of its website via Indiegogo.

"Crowdfunding allowed us to pool money from our friends and relatives as well as their connections, all under one website. There is a minimum cut that Indiegogo will take but overall it's worth it, it helped us build our first product," Mathieu Morel, Head of Business Development at MillionDC.

Difference Between Seed, Series A, B, and C, and IPO

Seed financing is the first step in getting money to start your business. It is also called startup financing. It could be your own money, a personal loan from credit company, crowdfunded, or from an angel investor.

As Investopedia explains, seed capital is the money raised to begin developing an idea for a business or a new product. It generally covers only the costs of creating a proposal that can be taken to venture capitalists in order to obtain additional financing.

Series funding usually comes in after rounds of seed funding when you already have proof of business and want to scale up. Series funds are mostly given by VCs. Series A, B, and C, are funding stages that are usually hinged on milestones that you set from proof-of-concept to operation, to revenue.

IPO or Initial Public Offering is when you sell your shares publicly, the stage when VCs and everyone who invests in your company becomes happier because they have a chance to cash in on their investment for huge gains. Going public has many requirements, one of which is consistent yearly earnings at a capped level.

Startup funding sources in Taiwan

As already mentioned before, the easiest way to get the money that you need to start your own company is through your savings and properties, and through your family and friends. Aside from these, there are many companies, groups, individuals and organizations in Taiwan that give funding to startup companies. They are the places where you can get bigger money that you need to start your company in Taiwan. Here are some of those companies:

1. Private Investors

There are two kinds of private investors in Taiwan that can give you the money to start your company. They are venture capitalists and angel investors.

i. Venture capitalists

Venture capitalists are companies or organizations that are really looking for startup companies where they could invest their money. They get the money by telling their investors to buy shares in the companies that they will finance.

If you accept funding from venture capitalists, they will require you to have their own people to take part in the running of your company. But that is good because their people have the ability to help your company to grow. But if you want to control your company, perhaps you should not choose venture capitalists.

ii. Angel investors

Angel investors are very rich persons who are looking to grow their money exponentially. They may ask for part ownership of your company the moment they give you the money or when it is already growing and earning well in the form of dividends.

Some angel investors form groups to invest together. They usually invest at the seed financing stage and so this could be right when your business is just trying to launch.

2. Taiwan government

The Taiwanese national and local government levels offer grants, incentives, and subsidies that can be taken advantage of by those seeking to start new businesses. There are three types of money that the Taiwanese government wants to give you:

i. Grant – this is the kind of money that the Taiwanese government will give to a company. But to get this grant, the government will check if it is right for them to give it to you. If the government gives you this grant, it will not expect you to pay it back. That means it is a gift given to you so that you can start your company in its country.

ii. Subsidy – this is money or cash that the government of Taiwan will give to you to pay for the money that you will spend in your startup company. For example, you need to pay your employees their salaries. You can get this subsidy to pay for the salaries of your employees. The subsidy can also be used to reduce your tax so that you will not pay big taxes to the Taiwanese government.

iii. Loan – this is money that the Taiwanese government will give you so that you can start your business in Taiwan. But you must pay back the amount to the government. The government will tell you how many months or years you have to pay back this money. You have to agree with this type of payment.

Most grants given by the government of Taiwan are called 'marching grants.' It means that you will also spend or invest your money on your startup company at the same time that the Taiwanese government is giving you this grant.

Extensive list of national and local Taiwan government funds for startups

The Small and Medium Enterprise Administration of The Ministry of Economic Affairs has listed a number of awards, grants and loans that the government provides from the national to local levels with the aim of encouraging SMEs or startups to start and run their business. Below are the lists culled from the MOEA website, with additional information directly from each of the funding source.

1. Taiwan SMEs Innovation Award

The Taiwan SMEs Innovation Award is aimed at promoting innovative R&D that allows for SMEs to prosper. For the past 25 years, the award has been granted to almost 860 SMEs that exhibited excellent performances in innovation and research development.

"*SMEs that persistently take an organized and systematic approach to engage in innovative research and achieve concrete results shall be awarded*," SMEA noted.

- Award amount: NTD 150,000
- Website: https://www.moeasmea.gov.tw/article-en-2612-4471
- Facebook: 中小企業創新研究獎 Taiwan SMEs Innovation Award
- Contact: 886-2-2366-0812 ext.308

2. Business Startup Award

Business Startup Award is aimed at encouraging startups to develop innovative products, technology, processes or services, and to create premium business models. Qualified enterprises are those less than five years old, and recognized by the competent authority.

"*For the past years, the award has been granted to 240 enterprises that exhibited excellent performances in Business models. The purpose of the award is to create more Startup company*," SMEA noted.

- Total Award amount: NTD 2,400,000
- Website:
    - https://startupaward.sme.gov.tw/Home/Index/#award/
    - https://www.moeasmea.gov.tw/article-en-2612-4472
- Facebook: 新創事業獎 Business Startup Award
- Contact: 886-2-2366-0812 ext.170 / 327

3. Service Industry Innovation Research (SIIR)

The Service Industry Innovation Research (服務業創新研發計畫) – Research grant for companies legally registered in Taiwan, not including branch offices, with positive equity in the retail, logistics, restaurants, advertising, e-commerce industries.

- Grant amount: Up to NTD 10 million
- Website: http://gcis.nat.gov.tw/neo-s/Web/default.aspx
- Contact: 886-2-2701-1769 ext.231 ~ 241

4. National Development Fund Startup Angel Project

The National Development Fund Startup Angel Project encourages entrepreneurs with innovative capacities and global potential to grow rapidly.

"*This Program is a free application. It has not entrusted any private business to host the charge seminars, or charge to write service of plan books and delivery services. Please do not believe that there are special relationships or channels! If in doubt, please contact the Investment Service Office of Business Angel Investment Program for confirmation (sic)*," NDF noted on its website.

- Investment amount: Up to NTD 20 million
- Website: https://www.angelinvestment.org.tw/
- Contact: 886-2-2546-5336

5. Taipei City Industry Incentive Subsidy Project

The Taipei City Industry Incentive Subsidy Project is aimed at encouraging technology development. Taipei is the first city government that provides R&D and brand establishment incubation, helping new startups and key competitive industries to develop. Companies registered in Taipei or SMEs are eligible to apply.

"*SMEs are qualified to apply for subsidies as long as the company is located in Taipei City with a new investment project which is innovative or has character or potential, there are no limitations of capital or industrial field*," The Taipei City Government said on its website.

- Subsidy amount: Up to NTD 5 million
- Website: https://www.industry-incentive.taipei/page-about-en.aspx
- Contact: Callers inside Taipei City 1999 (Outside of Taipei City 886-2-27208889) ext.1429, 1428

6. SME Innovation Development Project Loan

The SME Innovation Development Project Loan is for legally registered companies in Taiwan. The eligibility requirements for this loan are listed below:

i. The responsible person of the company is aged between 20-45 years old.

ii. The company has won an innovation prize from the government.

iii. The company has received subsidies for research purposes from the government.

iv. The company has cooperated with overseas companies, governments or academic institutions or has been commissioned by overseas, or has provided development services for research purposes.

v. The company has won an enhanced investment SMEs project from the government.

vi. The company has a research or production center in a national science park.

vii. The company has registered at the Taipei exchange center.

viii. The company has registered as a social innovation enterprise or has received B Lab certification by B Lab.

- Loan amount: Up to NTD 80 million for capital expenses and up to NTD 20 million for revolving funds
- Website: https://www.moeasmea.gov.tw/article-tw-2403-4074
- Contact: 0800-056-476

7. Small Enterprise Loan

The Small Enterprise Loan is for registered companies with less than 10 employees and is provided by private banks.

"SMEs must apply for loans through the banks because the purpose of the Small and Medium Business Credit Guarantee Fund is to provide the SMEs' credit guarantee and to ultimately share the banks' risks while encouraging the SMEs to seek loans," said MOESMEA on its website.

- Loan amount: Up to NTD 5 million for capital expenses and up to 80% of the project for revolving funds
- Website: https://www.moeasmea.gov.tw/article-tw-2403-4076
- Contact: 0800-056-476

8. Ministry of Education Universities Graduates Entrepreneur Program (U-START)

The Universities Graduates Entrepreneur Program (U-START) through the Youth Development Administration started in 2009 and has continued supporting new entrepreneurs in their endeavors. In order to qualify, the following requirements must be met:

I. Startup teams must have at least three members.

II. Two-thirds or more of the members must be students graduated from college or university in the past five years or current students (including undergraduates and postgraduates).

III. 18-35 years old non-students or foreign nationals holding a resident certificate are allowed to join the team.

IV. Each person may only join one team, and teams should be connected to an incubating college or university.

- Grant amount: Up to NT 1 million
- Website: https://www.yda.gov.tw/Content/QandA/contents.aspx?&SiteID=563655426603362361&MmmID=746301557642302131&SSize=10&MSID=2017120410270897234
- Contact: 886-2-7736-5111

Case interview: How do Taiwanese government grants/ subsidies work for startups?

We talked to Mark Hsu, a Taiwanese-American serial entrepreneur and investor, who is currently managing a USD 30 Million fund and he gave us a brief background on how grants and subsidies work in Taiwan.

The Taiwanese government has earmarked billions in USD to jumpstart Taiwanese startups and the money is parked at various government agencies and at the central and local level, he explained.

Hsu advised that startups think of this money as a subsidy. "You cannot apply for a 'blank check' grant based on a project and later decide on how you will spend the money. The majority of the grants in Taiwan are (actually) subsidies, meaning you need to spend the money first and after you have shown proof that you have spent the money, then the government will give you the subsidy per your grant proposal," he said.

First things first, register your company

Mr. Hsu noted that grants are given to companies and not individuals, and so the first step is to have a Taiwanese entity. If you are a foreign entrepreneur with a foreign company then what you need is to own a Taiwanese entity. Branch offices and representative offices are not allowed to apply for grants, he further noted.

Make sure you have the right amount of paid-in capital

"While your CPA may have told you that Taiwan no longer requires a minimum paid-in capital,...

The reality is that government grants are heavily tied to your paid-in capital. For example, the easiest grant is an NTD 1,000,000 startup grant offered by the Taipei City government but in order to receive that grant, you will need NTD 1,000,000 to start. In most situations, the grant you receive will be a discount to your paid-in capital," he further noted.

Record your expenses correctly:

Claiming for subsidy relies heavily on your receipts. Mr. Hsu said Taiwanese accounting is based on a framework that he calls "guilty until proven innocent." That is, unless you have the proper evidence, the expense cannot be properly recognized. "In the context of startups, what this means is that the money you spent on a SaaS [software as a service] subscription with your credit card to a company that is not registered in Taiwan is not considered a valid expense," Mr. Hsu noted.

Expense the right employee

The bulk of any government grant that you receive will be earmarked for salary subsidies. Mr. Hsu said foreign workers may be eligible for subsidies but only if they have Taiwanese residency (ARC). Payments to a remote worker will not qualify for a subsidy, he said.

Comprehensive list of VCs in Taiwan

VCs in Taiwan are always looking for companies that have the biggest chances of growing. Usually they will buy company shares and take part in running the activities of the company that they will give money to. They will do this by asking to have a seat in the company's Board of Directors.

Because they are investing the money of the company and not their own money, they are not very willing to invest in startup companies. But if you convince them that your startup company will really grow big and that you can give them a large profit on their investments, you may get the money you need.

Below is a comprehensive list of active VCs in Taiwan:

I. CDIB BioScience Ventures I, Inc.
10F., No. 225, Sec. 3, Beixin Rd., Xindian Dist., New Taipei City 231, Taiwan

http://www.senhwabiosciences.com/

II. Concord Venture Capital Group
4F., No. 76, Sec. 2, Dunhua S. Rd., Da'an Dist., Taipei City 10683, Taiwan

http://www.concord.com.tw/content.aspx?websn=47

III. First Venture Capital Co. Ltd.
9F., No.30, Sec. 1, Chongqing S. Rd., Zhongzheng Dist., Taipei City 10005, Taiwan

https://www.firstholding.com.tw/sites/fvc/overview

IV. Fortune Venture Investment Group
8F., No.16, Ln. 77, Xing'ai Rd., Neihu Dist., Taipei City 11494, Taiwan
http://www.vcfortune.com/

V. GAINS Investment Corp.
26F., No. 88, Chenggong 2nd Rd., Qianzhen Dist., Kaohsiung City 80661, Taiwan
http://www.gains.com.tw/index.html

VI. H & Q Asia Pacific
32F.-1, No. 333, Sec. 1, Keelung Rd., Xinyi Dist., Taipei City 11012, Taiwan
https://www.hqap.com/index.php

VII. Hoss Venture Inc.
1F., No. 36, Mingxian St., Sanmin Dist., Kaohsiung City 80794, Taiwan
http://hoss.com.tw/V/Vindex.htm

VIII. HQ Investments
30F., No. 213, Chaofu Rd., Xitun Dist., Taichung City 40757, Taiwan
http://www.hq-invests.com/#/

IX. IBF Venture Capital
8F.-6, No.188, Sec. 5, Nanjing E. Rd., Songshan Dist., Taipei City 10571, Taiwan
http://www.waterland-vc.com.tw/

X. Industrial Technology Investment Corporation
6F., No. 106, Sec. 2, Heping E. Rd., Da'an Dist., Taipei City 106, Taiwan
http://itic.com.tw/

XI. International Network Capital Management Corp.
10F-2 Ruentex Banking Tower 76 Tun Hua South Road Sec.2 Taipei Taiwan
https://wiharper.com/

XII. Maxima Capital Management
No. 16, Ln. 66, Sec. 4, Heping E. Rd., Wenshan Dist., Taipei City 11655, Taiwan
https://www.maximavc.com/index.html

XIII. Orion Venture Partners
http://www.orionventurepartners.com/

XIV. Power World Capital Management
8F., No. 70, Sec. 3, Nanjing E. Rd., Zhongshan Dist., Taipei City 10487, Taiwan
http://www.pwcm.com.tw/

XV. Premier Capital Management Corporation
10F., No. 76, Sec. 2, Dunhua S. Rd., Da’an Dist., Taipei City 106, Taiwan
http://www.premiervc.com.tw/default.htm

XVI. Shin Kong Venture Capital
38F., No. 66, Sec. 1, Zhongxiao W. Rd., Zhongzheng Dist., Taipei City 100, Taiwan
http://www.shinkonggroup.com/tw/sk_venture.html

XVII. Sunsino Venture Group
17F., No. 2, Ln. 150, Sec. 5, Xinyi Rd., Xinyi Dist., Taipei City 11059, Taiwan

https://www.sunsino.com.tw

XVIII. Vincera Capital
4F., No. 183, Sec. 1, Dunhua S. Rd., Da'an Dist., Taipei City 106, Taiwan

https://www.vinceracapital.com/home.html

XIX. Walden International Taiwan
18F.-2, No. 76, Sec. 2, Dunhua S. Rd., Da'an Dist., Taipei City 10683, Taiwan
http://www.waldenintl.com/index.aspx

XX. WK Technology Fund
10F., No.89, Sec. 2, Tiding Blvd., Neihu Dist., Taipei City 11493, Taiwan
http://www.wktechfund.com

XXI. Yuanta Venture Capital Co., Ltd.
10F., No.66, Sec. 1, Dunhua S. Rd., Songshan Dist., Taipei City 105, Taiwan
https://www.yuanta.com/VentureCapital/

Case study: Government accelerators and prototyping facilities

As the national government increasingly pours in resources to rebuild Taiwan as a tech powerhouse in the Asia-Pacific region, it is also playing catchup to its old glory days of being hailed as a destination for innovations. This means leveraging its 47-year old research institute, Industrial Technology Research Institute (ITRI) that is responsible for turning Taiwan's economy from labor-intensive to innovation-driven.

In this age of interconnectivity, more value is given to seamless fusion of hardware and software services. To be sure, Taiwan still has aces up its sleeve in hardware sector growth. However, the challenge has become greater at the onset of the 2020s as the coronavirus pandemic pressures innovators around the world to scour for solutions to keep people productive and healthy.

To solve these challenges, ITRI is on the lookout for innovations across industries and extensively offers the support of its own resources. The government agency offers its facilities to startups for prototyping and testing of their technology before releasing to the market. It also has an accelerator program to assist even individuals with their business ideas.

Prototyping and testing your ideas - Access to technical and electronic solutions

We interviewed ITRI Entrepreneur in Residence Jack H. Cheng who is in charge of global affairs within the IoT Integrated Service Center (IisC) at ITRI, where he helps foreign startups receive technical and electronic solutions through ITRI's engineering resources. IisC also offers business consultation and pilot production services to speed up product commercialization.

"Taiwan is a one-stop shop for global startup companies. Here you can find mature supply chains: system design service, electronic components and chip fabrication, manufacturing service with complete ecosystem, governmental incentives and supporting projects, language competency, etc. Even intellectual property will be available for licensing from renowned research institutions or universities," Cheng said, explaining why Taiwan is a great destination for foreign startups.

He said foreign-owned startups are able to acquire IisC's support at a minimal cost. Moving to Taiwan is not mandatory; foreigners can access IisC services such as technical evaluation by their engineers online.

Cheng noted that the goal of IisC is not to generate revenue from the startup partners, but to add value to their products. Although the service spectrum is wide, the IisC's objective is to optimize the existing product design with the consideration of cost structure, performance and even the protection of product design through the advanced manufacturing approaches which Taiwan significantly excels in among other countries. "We can help transform a great idea into a desirable product with the best quality and affordable cost structure," he noted.

Ultimately, he hopes that these efforts will entice foreign startups to shift their production to Taiwan.

Challenges faced by startups

One of the challenges for ensuring success of partnerships between IisC and startups is building trust, and Cheng highlighted that this is a two-way relationship. "Startups don't provide as much information, and I understand...this [the product design] is their baby," he noted.

He also observed that local VCs are conservative when investing in foreign hardware startups.

Having spent almost two decades in the US serving high-ranking IT managerial positions and corporate governance positions in local public companies, Cheng was recruited as the Entrepreneur in Residence at ITRI. "EIR's mission is to coach the engineering team to familiarize different business conduct, reposition product design or service model to address the pain points, establishing supply chain and pricing strategy, and mentoring fund raising technique and practice, hoping to help more ITRI spin-offs become promising enterprises."

When asked for his top criterion for identifying a good potential startup entering the IisC service platform, he said he looks for one who knows what the customer pains are, and provides a real solution to this pain with an actual product, or at least a prototype.

"Don't give up... As a founder of a startup, you need 100% support from your family so that you can focus on startup business. Being turned down 100 times is normal when seeking investors. Your founding partners are your most valuable asset, treat them well," Cheng noted.

Taipei's startup innovation cluster

There are other departments within ITRI that invest in startups that are founded by foreigners. In a separate interview, ITRI Business Director Thomas Chang said there are three criteria that he considers for investing in a startup namely team composition, business growth model, and barriers to entry.

Chang is responsible for startups coaching, business development, pre-A fund-raising, and preparing for listing at the Taiwan stock exchange. His specialties include Lighting, Robotics, IoT (internet of things), and medical devices.

Chang is currently in charge at t.Hub, the first build-order-transfer (BOT) facility in Taiwan aimed at promoting industrial development by nurturing innovation through startup projects.

The BOT project, a collaboration with Hongwell Group (宏匯集團), is located in Taipei's innovation cluster Neihu Technology Park (NTP). t.hub leverages key partnerships with ITRI and the advanced technology clusters of NTP to create a vibrant innovative startup community.

Chang's advice to foreigners who wish to start a business in Taiwan is to "find a suitable space, which can provide complete services for foreigners" such as the services provided by t.Hub.

He noted that t.Hub has a high occupancy rate, given that it is in the middle of a technology park. For instance, they have already contracted a number of big companies such as LINE.

Case study: Tracing the illusive VC funds

Foreign-owned startups are struggling with raising funds in Taiwan and this problem can be addressed by having a well-coordinated effort across sectors that would support startups at different stages of growth, according to Viktor Berglind, Senior Advisor at Abso Capital.

"Many startup companies I meet in Taiwan, are struggling on the funding side. They find it difficult to connect with venture capitals. There seems to be less than a dozen VC companies that are active and growing that [number] is important... There is certainly a need to have a more dynamic fund-raising climate here in Taiwan. We need to see coordinated efforts across academic, public, private, and government bodies to nurture companies and support them throughout different growth stages," Berglind said in an interview with MillionDC.

He said that VCs in Taiwan tend to look at later stage companies, typically Series A and onwards, those with a proven product and some revenue who need funds to grow and take their businesses to the next level. This, when many of those who need funds are still in the very early stage of product testing and prototyping. He also noted that outside Taiwan, many VCs cover a broader spectrum of the growth cycle, with both VCs and family offices being more open to early-stage and seed funding.

"You would find a lot of companies who need capital to get their products going, for product testing and prototyping, looking at seed rounds. These types of funds usually come from friends and family but not everyone has access to friends and family who are willing to write a check to support a crazy idea," Berglind explained.

He said that access to family offices is quite a challenge partly because of the language barrier but largely also because of the networking aspect referred to as guanxi. "In mature markets, family offices with a longer operating history would be more prone to a cold email. But that's a relatively measurable [outcome]," he said.

He said another option for funding would be the assistance of incubators, but they take a stake in a startup, which may be fair, but this is not a suitable model for everyone.

"Increasing the pool of available capital is important to build a healthy startup ecosystem...it is an effort that requires broader support," Berglind noted.

For his work at Abso Capital, Berglind explained that he is interested in helping those startups that are tackling sustainability, aligned with the 17 Sustainable Development Goals listed by the United Nations.

"We are looking at projects that have interesting growth prospects and also have a social good tied to it. Personally, I want to be able to tell my son that his father, in a humble way, made this world a better place," Berglind said.

Commenting further on his recommended policy changes for Taiwan's startup ecosystem in general, Berglind highlighted the need for support across sectors. For example, he recommended identifying what certain cities or regions are good at and providing support by directing the right resources tailored for that location.

"Not every district can be Silicon Valley. Taiwan cannot have a single industry guiding the whole country... it has to be in a number of sectors," he noted.

Come to Taiwan to serve the Taiwanese market first

In another interview with a managing partner at a Taiwanese VC who allowed to be interviewed on the condition of anonymity said that every foreigner coming to set up a company in Taiwan has to be clear with his intentions to open a business here and it should be to primarily cater to the Taiwanese market.

"Once you build a business here, the first challenge is that we have to localize, test your business model, if it will attract local people. If you cannot speak Chinese, you can't mingle locally, you're facing a lot of challenges," the VC said.

He noted that over the years, in terms of regulations, the government has put a lot of incentives for foreigners to come to Taiwan to build their own business.

"I think in the country point of view, we would like to be more diversified...similar to, for example, in Bangkok you can see a lot of foreigners walking on the streets, building their own businesses and make their culture mingle with the locals." he said.

"Things have changed a lot, accelerators here don't think a lot about if you are a foreigner or Taiwanese. They only look at your business model," he noted.

Taiwan is a manufacturing- and biotech-biased economy

"The only problem for foreigners is that Taiwan's economy is very "biased" [in] that it is very focused on high-tech manufacturing and also biotechnology. Outside these two major fields, it would be hard to set up a business," the VC noted.

Because of this bias, foreigners who wish to set up a business in Taiwan, outside its economic bias must think of expanding from Taiwan. "For example, if you want to build a great social media empire, you have to do cross-border," he said.

Jumping board to Southeast Asia and China

Foreigners setting up business in Taiwan also benefit from its close proximities to China. "We are very close [similar] to China, in terms of culture, in terms of the history... Compared with Hong Kong, our living standard is quite cheap... Moving your family over here, you won't feel any pressure," the VC noted.

He said one of the advantages of Taiwan is that information is transparent and the government will not limit you to access vital information for starting a business.

"Taiwan is a very good hub, you can use Taiwan as a [jumping board] to Southeast Asia and China market," he said.

Capital restrictions

He said the lack of capital freedom in Taiwan, with the government requiring clearance before any foreigner investor can wire capital in and out of the country, stops foreign VCs from coming. This in turn makes it less attractive to foreigners to set up businesses here compared to Hong Kong and Singapore, where VCs are able to easily wire money in and out of the country as they inject capital into different markets.

"Money laundering is one issue, exchange rate [control] is another issue... That is why if you come to Taiwan, there are not so many international private equity firms nor venture capitals. In Hong Kong, most international private equity investors set up headquarters in Hong Kong because [they] can easily inject US dollar investment capital inside or to Taiwan or to China," the VC noted.

"I used to work for investment banking for 15 years and I understand the regulations in terms of capital injections. Taiwan could remove all the capital limitations like Hong Kong, so that capital floods to Taiwan and once foreign VCs set up, more foreigners [startups] will come," he noted.

In a recent development, President Tsai Ing-wen told the Taiwan Capital Market Forum in August 2020 that Taiwan would liberalize banking and investment rules to establish itself as a regional financial hub.

Taiwan once held the highest VC density title

The VC said Taiwan used to have the highest VC density within one country in the world. This "was back in the 1980s [with] most of the big companies like Acer, ASUS, TSMC. [Back then] if you cover your eyes and invest in any kind of high-tech manufacturing, you'd go IPO afterwards." he noted.

He said that Taiwan didn't transition very well as 80% of the IPO market is dominated by high-tech manufacturing. This, when the world is focusing less on this industry.

"[The number of VCs] is also related to the value of the IPO market. For example, we haven't seen a Taiwanese startup listing in NASDAQ in the last 20 years. But in China, every month, there is one company listing in the NASDAQ. Most private equity and VCs set up a company in China but they don't set up here in Taiwan," he noted.

When asked about an upcoming VC that claims to be accessible worldwide from its lone office in Indonesia, he said "VC is a local business, just like a startup...

If you want to come to Taiwan, you have to treat yourself as a local business to understand your market. [You] will need to put up a lot of resources to understand what this startup does, what it has done in Taiwan, and how the business model works...

"It doesn't make sense to say 'I am coming to Taiwan' and targeting another country," he said.

Criteria for investing

As VC, his investment focus is on early stage startups who are looking at expanding to the greater China region and Southeast Asia.

"We're doing early stage, so the people [team] is very important and how they are trying to execute their business plan. If you tell me you are a single guy [one-man team] and how mighty you are, I will not believe it because you know a single guy would have a lot of drawbacks, a lot of areas that you cannot cover," he noted.

He said he'd also invest in ideas even on tissue paper "If it is only just an idea I will put 100% [of judgement] on the team. My philosophy is very simple. If I invest in an early stage [startup, I] won't pay attention about what kind of project they want to do. I care about the team. Even if they fail on a project and there is money, they can still make something out of it and the team still belongs to me," he noted.

Unlike a normal VC that invests in six companies in a month, his firm only invests in five to six per year because "we're focusing on whether we can add value to a company... If I find out that my value to a startup is just money, then I won't invest in it."

Foreign startups sentiment on Taiwan VCs

MillionDC conducted an informal survey about the overall experience of foreigners with VCs in Taiwan. We conducted an open-ended question in two startup groups, a LINE group called Taiwan Startup Community, and a Facebook group called Foreigners Society in Taiwan.

No one claimed to have gone through a successful funding with VCs but some were accepted at accelerators. We decided to mask the identities of the startup founders.

Taiwanese VCs rely on guanxi

One of the respondents highlighted the "lack of openness" of Taiwanese VCs even when his startup would match the VC's field of interest and generate revenues.

"I think some of the improvements needed for VCs in Taiwan include being open to hear what exactly companies [startups] are doing... [For example] they say on their websites that they invest in fintech, deep tech and AI but when a fintech/deep tech company seeking for Series A even with a million-dollar revenue in the past approaches them, some of them are not inclined to hear [the pitch]. We later on found out that some startups who have built a prior good relationship with the VCs are the ones who are able to follow up," a fintech startup founder recalled.

"Some are friendly and some are arrogant... Many Taiwanese VCs rely on guanxis. The success rate of those who cold-pitch to Taiwanese VCs [is] lower than those who get referred," he noted.

Another startup founder however said that this sentiment is not unique to Taiwan.

"This is also true in Silicon Valley. If you have previous history, you get meetings easier, even if you failed [before]. The issue in Taiwan is most VCs are unable to do due diligence so you'd need a lead VC from Singapore or USA who can do the vetting [for your startup]," he said.

Why stay in Taiwan then?

The fintech startup founder noted that while they do not have a Taiwanese VC supporting them, they intend to stay in Taiwan to "expand and to recruit talents" noting that "talents in Taiwan are pretty resourceful, especially tech talents." He also said that they are backed by VCs from abroad.

Another founder noted that the above case could be isolated and claimed that he knows "personally some who got funding without [going through] guanxi, whose startups are in IoT, blockchain, and live streaming as well."

Chapter 5

How to Legally Reside in Taiwan as a Foreign Entrepreneur

Chapter Snippets:

For foreigners who are outside Taiwan or for those who want to have a more permanent resident status, this chapter provides a step-by-step process in getting either an entrepreneur visa or an Alien Residency Certificate (ARC). We begin by learning who are eligible to apply for an entrepreneur visa. This is followed by the step-by-step procedure in applying for an ARC as derived from a case interview with an agency who has helped many foreigners apply for ARCs. The second case interview is an actual experience of a foreigner who applied for an entrepreneur visa and how he was in visa limbo amid the Coronavirus pandemic.

Registering your company and applying for a legal way to reside in Taiwan are two different steps. You may apply for an entrepreneur visa or an ARC card. In general, an entrepreneur visa is loosely tied to the business that you're starting. Meanwhile, an ARC application requires you to obtain a work permit under your own company, which must meet a capped investment and turnover from the first year of activity. This is to say that an entrepreneur visa is usually the easier route for legally residing in Taiwan as a foreign business owner, in the short run. In the long run, registering your business, even a representative office would be the smarter move.

Who can apply for an entrepreneur visa?

1. Individual application

If you are the only one operating your business, you can apply for an entrepreneur visa in Taiwan. You can apply for an individual application under the following conditions:

I. You have money amounting to USD 63,600 or NTD 2 million. A venture capitalist or a crowdfunding international company may have given the money to you as an investment.

II. You are working on your own startup or are a member of a national startup business.

III. You have an invention or design that you have already registered with national or international authorities. Your own invention or design is called an intellectual property.

IV. You have won a big competition for a startup business or design, or have been given a prize at big film festivals.

V. The Taiwan government has already given you money called grants.

VI. You have already started a special business that is recognized by the Taiwan government. Such a business should have an investment of USD 32,000 or NTD 1 million, and you must own or be in charge of the company.

2. Team application

If you are composed of two or three people, you can also apply for an entrepreneur visa in Taiwan. This application is called team application. Your team could have a maximum of three members. For team application, you must show the Taiwanese government the following:

i. If your company is registered in Taiwan:

- You and your team members must show that you have already started a special company that the Taiwanese government has given its approval to.
- You must also show that you and your team members are the persons in charge of the company such as managers, directors or supervisors.
- Your company must have an investment of about USD 32,000 or NTD 1 million

ii. If your company is not registered in Taiwan, you and your team members should show the Taiwanese government that you have one of the following:

- You have money totaling USD 63,600 or NTD 2 million. A crowdfunding international company or a venture capitalist has given money to you as their investment. A crowdfunding company will give you money that comes from a crowd of people on the internet. They want to use their money to earn profit.

- You are a member of a national startup or are working on your own startup.

- You have a design, creation or invention which you have already registered with national or international authorities. Your own creation or invention is called intellectual property.

- You have won a big competition for a startup business or design, or have been given a prize at big film festivals.

- The Taiwanese government has given you a grant or a sum of money for something that you want to do.

How can you apply for an entrepreneur visa?

If you are already in Taiwan as a visitor, you can apply for the entrepreneur visa within 45 days before your visitor's visa expires. You can submit your application to the Bureau of Consular Affairs. You can also submit your application in the offices of BOCA near the place where you are staying. These BOCA offices are in Eastern Taiwan, Southern Taiwan, Southwestern Taiwan and Central Taiwan.

If you are not in Taiwan, you can go to the Taiwanese Embassy or the Taiwan overseas mission in your country and submit your application there. If you are in Macau or Hong Kong, you can submit your application to the National Immigration Agency or the Taipei Economic and Cultural Office.

If you will submit your application in Taiwan, it will take about two to three weeks before your application will be approved. It will take more than two to three weeks if you submit your application in a foreign country. The reason is that you need to send documents and write letters to Taiwan and you need to wait for the answers to your letters.

When you submit your application, your papers or documents will go through this process:

Step 1 - If you submit your application in Taiwan, it will be handled by the National Immigration Agency. But if you submit it outside of Taiwan, it will be handled by the Taiwan Embassy, or the Bureau of Hong Kong or Macau Affairs or the Taipei Economic and Cultural Office.

Step 2 - Your application will be checked by the Investment Commission. If some documents are not correct, you will be asked to correct them.

Step 3 - If there is no problem with your papers of documents, your application will be given to the National Immigration Agency. This agency will recommend the final decision on your application. They will do one of two things: a) approve your application or b) disapprove your application.

Step 4 - If the NIA approves your application, you will be given your entrepreneur visa. That means you can already start your business in Taiwan.

Step 5 - If the NIA disapproves your application, you will not be given an entrepreneur visa. That means you are not allowed to start a business in Taiwan.

Step 6 - If you are given an entrepreneur visa, you need to acquire the Alien Resident Certificate from the National Immigration Agency within 15 days of your arrival in Taiwan.

Documents needed to apply for an entrepreneur visa

There are several papers or documents needed in applying to get an entrepreneur visa:

i. Visa application Form

   a. You can download this from the internet. You need to visit this website: http://www.boca.gov.tw/cp-166-277-41131-2.html. Fill up this online form. Put in all the correct information about yourself and your business and follow the instructions on how to submit it.

ii. Passport and one photocopy of it

iii. 2x photos, colored and passport size

iv. Application categories form. You can get this form from this website: https://www.moeaic.gov.tw/businessPub.view?lang=en&op_id_one=6

v. Requested Documents (by the immigration)

vi. Letter of Intent – this is optional. You can visit this website to get a sample of this letter of intent: https://www.moeaic.gov.tw/businessPub.view?lang=en&op_id_one=6)

Case interview: Agencies need to catch up

Private agencies in Taiwan need to shape up in order to facilitate a growing number of foreign-owned companies who are looking to set up a business in the country, according to Fastart founder Carter Lee. Fastart is responsible for successfully helping hundreds of foreign-owned companies in Taiwan.

To be sure, the MOEA announced that 943 foreign direct investment (FDI) projects with a total amount of USD 2,422,103,000 were approved from January to March 2020. This involves an increase of 13.48% in the number of cases, and an increase of 136.44% in FDI amount compared to the same period in 2019.

As Taiwan gains attention worldwide, especially with the government's notable efforts in containing the number of 2019 coronavirus cases in a country that is only a few miles away from the virus' origin, more foreigners are expected to consider Taiwan as a place for opening up new businesses.

"*I have observed that there are still many agencies who are unaware of the new rules related to foreigners who want to set up business here. Our clients will usually do their research and ask other agencies, and then they will be surprised to know that the information they gather are not all correct,*" Lee explained.

Some accounting firms are not even open to entertaining foreign clients, based on MillionDC's experience when we were registering as a formal business in Taiwan. Lee explained that this could be because of the language gap, with some Taiwanese accountants uncomfortable using English.

"The younger accountants are more comfortable with English language and so they are mostly the ones who are able to help foreigners," Lee noted.

Foreseeing more and more foreigners willing to settle in Taiwan, he tries to handle as many cases as possible, to familiarize himself with the newer rules. "That's the only way to grow your competitive advantage, to take the harder route," Lee said. Because of this, he could take pride that his firm is able to provide service to clients quickly and on time. Lee said the first step his firm takes when hired by foreign companies is to gather information about each company's mission statement, operations, and business processes. From this, he can identify at first glance which industry rules a company belongs to.

Lee's specialization is in assisting foreign entities in setting up their company, applying work permits, and applying for ARCs, as well as clerical concerns such as bookkeeping and filing for taxes. Lee said foreign investors can apply for ARC through the following:

- One manager can qualify for a work permit to get an ARC by having a capital of NTD 500,000.
- One investment ARC and one manager work permit can be applied for with an investment capital of NTD 6,000,000.
- An investment capital of NTD 15,000,000 is treated as an investment immigration which reduces the necessary residency for APRC applications to three years.

Although these are the general guidelines, there are some special restrictions for investors of certain lines of business. For instance:

- If the business is centered on Tourism, a capital of NTD 1,200,000 is required.
- For businesses in the shipping industry, a capital of NTD 25,000,000 is required.
- On a case-by-case basis, there may be considerations imposed by the Taiwanese government.

Beyond these requirements, applying for ARC for foreign investors would have to be done with the assistance of a Taiwanese entity as proxy. A foreign investor going through Fastart's process will be assisted for the following steps:

- Name check
- Notarization
- Investment check
- Opening of bank account (personal)
- Company setup
- Taxation identity check (personal)
- Applying for a work permit
- Applying for an ARC (personal)

Unlike other companies with a similar line of business, Lee said that his firm also assists with the necessary legal paperwork and steps to set up the company locally. This includes accompaniment and assistance with the aforementioned investment checking, opening of a bank account, and preparation of business plans that follow the Taiwanese national and local government laws. Essentially, Fastart serves as the proxy requirement as part of their service package.

According to Fastart, the application process will have the following timeframe:

- 20 days to apply for a work permit
- 30 days to set up the company

After acquiring a work permit, applying for an ARC becomes easier. Fastart does note that the holder of an ARC must adhere to the limits of what is allowed. For instance, the ARC holder cannot work for another company besides the one stated in the work permit, or their ARC may be revoked and cannot be renewed for three years.

Although there are several ways of acquiring an ARC for a foreign investor, such as being a specialist hired by another company, or being employed by a foreign company and serving as that company's representative in Taiwan, Lee states the fastest and most effective way to acquire an ARC (and eventually an APRC) is to set up a company and fulfill the minimum required residency in Taiwan for the stated period.

"Do you know that other agencies, including banks, will ask foreigners who are already ARC/APRC holders to route your money from your home country and transfer it back to Taiwan? This case may no longer be necessary... This is just one of example of how many are not aware of the new rules, and having assisted many foreigners allows us to save time," Lee noted.

For assistance with registering your business, contact Carter Lee.

Case interview: Entrepreneur visa limbo

A: foreigner establishing his startup in Taiwan was left in visa limbo as Taiwan implemented stricter border rules due to the 2019 coronavirus pandemic. This case interview is left in Q&A form to preserve the context of answers of our interviewee who requested not to be named.

Q: Why apply for an entrepreneur visa? Could you continuously work on your startup without it?

A: The entrepreneur visa is an attractive option as it grants me residency and the ability to focus on my startup without needing to take the occasional trip out of the country. I could work on my startup without it (and have been doing so for some time) but it seems like a useful step to take as I exit stealth mode and move toward launch.

Q: What makes Taiwan an attractive place to start your business?

A: For me it's all about the quality of life in Taiwan. There are probably better places to launch from a business perspective, but I happen to really enjoy living here. Just wandering around the streets puts me in a good mood and keeps me feeling creative and inspired. If I need a break from slinging code in the city it's easy to go hiking, visit hot springs, or hop on a train and explore any number of other interesting places all around the country.

Q: Could you briefly give us a background on your startup, without giving hint of who you are (to protect your identity).

A: My startup is essentially a digital content distribution system focused on serving a niche market. Most startups in this space have taken VC funding and seem intent on mass adoption and ad-based business models whereas I am more interested in providing premium services to industry professionals.

Q: Could you explain your "in limbo" situation, related to applying for an entrepreneur visa?

A: I've been making regular trips out of the country to renew my visa-exempt status while preparing to apply for the entrepreneur visa. From what I've heard it isn't difficult to obtain the entrepreneur visa in the first place, but the requirements for renewal are difficult to meet for new startups, so it made sense to delay my application until I was ready to launch.

I was due for another visa run when the coronavirus pandemic began to disrupt international travel so I opted to convert to a 180-day visitor visa, just to be on the safe side. As soon as it became apparent the pandemic was not going away I went ahead with the entrepreneur visa application only to be told that I'd have to leave and return to Taiwan to begin the process. This was already impossible at that point, and various other options (like applying at a foreign office and securing a special entry permit to return) also wouldn't work. None of the extensions generously provided by the Taiwanese government to travelers stuck here during the pandemic applied to me as I already had 180 days on my visa.

With no better idea of what to do I hunkered down and continued work on my startup throughout March, April, and May.

Only in early June did the relevant authorities relax their policies about converting visas in-country, and I'm fortunate that I visited BOCA one last time to plead my case as they still haven't made any public announcement about this to my knowledge! My application was accepted one business day before my visitor visa expired.

Q: What could have been in place to avoid this struggle from happening, for others who may go through the same experience?

A: Clearer communication would have been helpful in my case. If someone had told me "You won't be able to convert this visa to any other kind without leaving the country" before I applied for it I would have made other plans. The extended visitor visa is marked "non-extendible" but I was intending to apply for a different kind of visa, and did not realize this qualifier would normally make that impossible. Ultimately this was my mistake, but there is little information available online, and the counter staff at the visa office, although friendly, were focused entirely on the task at hand, and did not clearly answer my questions about future plans. More flexibility in decision-making due to the pandemic would also be useful, of course. I was trying my best to make responsible decisions and landed in three months of bureaucratic limbo for my trouble.

Q: What could be the bottleneck in the current process, now that we are in pandemic?

A: International travel is the obvious bottleneck in the midst of the coronavirus pandemic. Taiwan's containment efforts are among the best in the world and there is obviously a strong desire to ensure the country remains free of the virus.

On the other hand, it will be difficult or even impossible to recruit foreign talent and startups as long as these restrictions remain in place, and there seems to be little interest in actively recruiting those skilled professionals who, for one reason or another, are already in the country and figuring out what to do next.

Q: Does the situation outweigh the attractiveness of Taiwan as a place for starting a business?

A: Enduring months of uncertainty wasn't pleasant, but I grew to accept the situation at some point, and it allowed me some extra time to work on my startup. If I weren't bootstrapping as a solo founder I would be much less enthused about the experience.

Q: Do you think there is also a shortcoming from your end? (To be fair with the article treatment) A: I probably should have asked more questions at BOCA and insisted on clear answers before filing paperwork. As noted, it was ultimately my mistake that I converted to a dead-end visitor visa.

Chapter 6

Soft-Landing Package Unpacked

Chapter Snippets

The first time living in Taiwan as a foreign business owner could be quite fun and challenging. If you're currently outside Taiwan, you may want to avail of a soft-landing package offering that includes not just setting up your business but also your permit and place for residence. We have also included an overall view of Taiwan business culture as well as do's and don'ts that every foreigner must learn in order to blend in smoothly as you scale up your startup here.

As the government tries its best to attract foreigners to set up businesses in Taiwan, many agencies have permits from the government to offer soft-landing packages to help foreigners legally stay and test their businesses within the island. These packages offer processing of necessary permits, visas, office space, and even living accommodations for a few days up to several months. Fees vary according to the scope. These services are especially targeted to those foreigners who are abroad.

iiiNNO Taiwan is one of the licensed agencies that can process soft-landing packages for foreigners who wish to set up a business in Taiwan.

One of the benefits of going through a certified agency for offering soft-landing packages is the speed of processing of papers. "We can guarantee fast track four to six weeks startup VISA and ARC application approval by the Taiwan government as we are one of few certified accelerators / incubators in Taiwan," said iiiNNO founder David Kuo in an interview.

Another perk is providing consulting for accessing research and development grants. Kuo said their company has helped a company apply for a government subsidy. For example, one of iiiNNO's government partners is the Small Medium Enterprise Administration (SMEA). "Under SMEA's annual subsidy, we are able to focus on helping very early-stage entrepreneurs from idea to startup project with possibly creation of an actual startup company," Kuo noted.

Below is sample of soft-landing package offer:

BASIC PACKAGE WITH SUBSIDIARY COMPANY SETUP	ADVANCED PACKAGE WITH LAUNCHPAD SERVICE	PREMIUM PACKAGE WITH LAUNCHPAD & SHARED SERVICE
3 MONTHS FAST TRACK TAIWAN SUBSIDIARY COMPANY SETUP	3 MONTHS FAST TRACK TAIWAN SUBSIDIARY COMPANY SETUP	3 MONTHS FAST TRACK TAIWAN SUBSIDIARY COMPANY SETUP
included: 3 months English-Chinese Accountant Service	included: 4 months English-Chinese Accountant Service	included: 4 months English-Chinese Accountant Service
included: 1 time startup visa application	included 1 time startup visa application	included: 2-time startup visa application
included: 1-time ARC (residence card) application	included 1-time ARC (residence card) application	included: 2 time ARC (residence card)
included: bank account application and setup	included: bank account application and setup	included: bank account application and setup
6 MONTHS ADDRESS & MAIL SERVICE SERVICE	6 MONTHS ADDRESS & MAIL SERVICE SERVICE	12 MONTHS ADDRESS & MAIL SERVICE SERVICE
2 HOURS *UP-TO 150K USD* GOVERNMENT RD SUBSIDY APPLICATION CONSULTING SESSION	4 HOURS *UP-TO 150K USD* GOVERNMENT RD SUBSIDY APPLICATION CONSULTING SESSION	6 HOURS *UP-TO 150K USD* GOVERNMENT RD SUBSIDY APPLICATION CONSULTING SESSION
–	1 TIME DRAFT VERSION REVIEW	2 TIME DRAFT VERSION REVIEW
3 MONTHS 7/24 CO-WORK SPACE ACCESS (1 HOT SEAT)	4 MONTHS 7/24 CO-WORK SPACE ACCESS (2 HOT SEAT)	5 MONTHS 7/24 CO-WORK SPACE ACCESS (3 HOT SEATS)
included: overnight shower service	included: overnight shower service	included: overnight shower service
–	–	1 FREE MONTH DEDICATED 1~2 PEOPLE OFFICE
–	LAUNCHED PAD: LAWYER/ ACCOUNTANT REFEERAL	LAUNCHED PAD: LAWYER/ ACCOUNTANT REFEERAL

–	LAUNCHED PAD: 1~2 SUPER CONNECTOR MATCH	LAUNCHED PAD: 2~3 SUPER CONNECTOR MATCH
SHARED SERVICE: MAIL SERVICE	SHARED SERVICE: MAIL+BANK	SHARED SERVICE: MAIL+BANK
SHARED SERVICE: RECRUITING POSTING	SHARED SERVICE: RECRUITING POSTING & INTERVIEW SETUP (1~2 CANDIDATE)	SHARED SERVICE: RECRUITING & INTERVIEW SETUP (1~2 POSITION WITH 3~4 CANDIDATES)
–	–	SHARED SERVICE: TALENT PRE-SCREENING (1~2 INTERVIEWS)
startup company has a strong desire to complete the Taiwan subsidiary company setup within 3~4 months. Moreover, would like to leverage iiiNNO experience and network for government subsidy application and talent acquisition.	startup company has a strong desire to complete the Taiwan subsidiary company setup within 3~4 months and launch service within 4~6 months. Moreover, would like to set up an operation/RD/sales team to leverage iiiNNO experience and network for government subsidy application/talent acquisition/building local network. With iiiNNO shared service team, the effort of building local operations can be minimized or delayed.	startup company has a strong desire to complete the Taiwan subsidiary company setup within 3~4 months and launch services within 4~6 months. Moreover, would like to leverage iiiNNO years experience of helping international companies not only to set up an operation/RD/sales team, but successfully launch service with possible local customization in mind. With more extensive shared service, iiiNNO team will collaborate with your team in government subsidy application/talent acquisition/building local network.

Before landing: Study Taiwan business culture

On your first days in Taiwan, or even before you land, it is important to learn about Taiwanese business culture. Taiwanese business culture fully embraced capitalism in the 1980s, leading to the emergence of an entrepreneurial spirit that we know it today.

In more established businesses, the organizational structure is hierarchical. For example, it is considered polite to greet or address the oldest or the most senior person at meetings. While there is conscious effort among new companies especially in the tech industry to change, decisions are usually made from the top down and not questioned by subordinates. One very important trait is that the Taiwanese value long-term friendships in business to ensure that they can count on their foreign counterparts. Below are some useful tips to follow Taiwan's business culture:

- Dressing for business

  Taiwan has a similar business dress code with the US; formal and conservative for men and women. It is advised that men wear dark colored suits with shirts and ties and women wear conservative business suits or dresses and blouses. Men usually take off their jackets during meetings.

- Shoes

  Wear leather shoes to match suit attire. Always take your shoes off when you enter someone's home. This applies to all of the homes in Taiwan. Hosts usually provide slippers.

- Greetings

  Shake hands with a nod of your head. The Taiwanese don't bow as Japanese people do. In a more formal setting, men and women simply nod and men should wait for a woman to initiate a handshake. Gifts are offered in both hands.

- Business cards

  Offer/receive business cards with both hands. Do not store the business card that you receive right away. Taking time to read the card's detail is a sign of respect. The cards should remain on the table in front of you, during the entire meeting.

- Meetings

  Saving face (avoiding embarrassment) is an important aspect of business interactions during meetings. Because of this, the Taiwanese avoid saying no and they may tend to give vague responses. Therefore, non-verbal cues are very important when evaluating their response. When presenting, speak directly to the most senior person in the meeting. Don't use red ink when writing someone a personal note. Red ink is associated with protests, denunciations, and corrected exams.

- Dining

  Business meetings are usually followed by a long dinner. The Taiwanese use chopsticks, so give it a try. Don't play with your chopsticks. Do not leave them stuck to your bowl of rice because this represents a cursing of the host similar to an offered prayer in shrines. Pick up your bowl of rice close to your chin when eating. Don't discuss business unless the Taiwanese team brings it up. Tea is served at the end of the dinner. When someone pours a drink, it is polite to say thank you by gently knocking on the table next to your glass with your index and middle fingers, representing a kneeling person. It is ideal to reciprocate the dinner of the same value.

- Guanxi

  Many businesses in Taiwan are built on guanxi, which simply means relations. To strike a business means nurturing and strengthening relationships over time. Establishing guanxi is more important than striking your first deal as the Taiwanese tend to extend the business relationship to their network solely on bonding.

- Gifts

  Don't open a gift in front of the person who gives it to you. Don't give an umbrella as a gift or company giveaway. The Chinese word for umbrella, "san" sounds like the word for "to break apart." Don't give a clock as a gift or company giveaway. The phrase "to give a clock" (sung chung) sounds just like "to attend a funeral": it's very inauspicious. Don't give a handkerchief as a gift or company giveaway. In any case that you accidentally bring an "unlucky gift," this can be settled by allowing the receiver to give a coin to make it a purchased item.

The above summary was culled from the following sites

https://santandertrade.com/en/portal/establish-overseas/taiwan/business-practices

https://www.roc-taiwan.org/usnyc_en/post/132.html

Chapter 7

Case Studies

Chapter Snippets:

To further illustrate the procedures and processes involved for foreign companies that are being registered in Taiwan, we took time to interview some founders on their experience. The first case deals with an internationally acclaimed advertising agency that shares how their business started rolling partly from government subsidy. The second case is about a half-Taiwanese businessman's startup that has been awarded many grants abroad but not yet in Taiwan. The third case involves the matter of tax rules that are a bit confusing for foreigners. Finally, the fourth case is a company that has been successful in applying for not just one but three accelerator programs in Taiwan, all in just 12 months.

Case A: How an ad agency got a government subsidy

BeLucky is a creative agency that has its main headquarters in Taipei while operating satellite offices in London and Prague. The agency has just been given a Platinum Award by the MUSE Creative Awards 2020 in New York, U.S.A. The agency had its humble beginnings in Taiwan way back in 2013.

In an interview with its founder and owner Karel Picha, from the Czech Republic, the first struggle with registering his business in Taiwan was with the close to zero translations of websites. Thanks in large part to his girlfriend who is a Taiwanese native and her mother, Picha was able to gather the necessary data and information in applying for his residency and in the legal procedures for establishing his company in Taiwan. This included applying for the NTD 1-million grant that subsidized his startup.

Picha said filing documents was necessary, but not very convenient. As the process could not be done online, everything was done in person, going to and from the offices in different locations.

"*Taiwan is one of the top high-tech countries in the world, yet most of the government and banks' related procedures are dealt 'offline' on paper. So, one of the biggest challenges would be Taiwan bureaucracy - where it requires quite a big amount of time and endless filling in all sorts of paperwork. And having to hand it in person in several different places*," Picha said.

He also recalled that there were also difficulties encountered due to the language barrier, as all the documents, including the documents for foreign investors, have to be both in English and Mandarin Chinese.

He said part of this difficulty was also speaking with the Taiwanese staff that spoke little or no English.

"*There is a special government department whose only purpose is to serve foreign investors and businessmen who want to open business in Taiwan. Every step of your business registration needs to get an approval from this department. I remember I had to go there around three times. Strangely enough none of the staff could speak English at that time. Once, they even got annoyed with me asking me why I don't ask some Taiwanese to go to deal with it. I like to believe it has changed since then*," Picha said.

Tips for getting a government subsidy

BeLucky Taipei applied for a subsidy from the Taipei government. Picha went through the regular review of documents and presentations and after learning that his agency had been approved, he learned one part of his business plan gave them the competitive edge for the subsidy.

"*The government appreciates when you are willing to give back to society, especially for Taiwanese. We didn't know that but part of my business plan was a cafe/art club below our advertising office. The cafe has been a platform for people to meet up and share all sorts of experiences (music, art, culture, food)*," Picha said.

At the time of BeLucky's application for a Taipei subsidy, the company only had NTD 30,000 as its paid-in capital, according to Picha. This case proves that foreign startups don't need to put in much paid-in capital in order to increase their chances of getting government grants. "They look at your plan," Picha noted.

NOTE: The government recently revised its guidelines for applying for the branding subsidy. Companies applying for this incentive must have their companies officially registered and running for at least one year with proven revenue stream.

Case B: How a recognized startup abroad was rejected thrice for a Taipei subsidy

One would think that a startup with a number of recognitions received abroad would just have a walk in the park when applying for a government grant. After all, most startups look for the same things: a product that solves a problem, and a market for that product. But apparently this wasn't enough for this company, whose founder requested anonymity.

His company specializes in the development of artificial intelligence and machine learning, including the application of such technology on financial markets, as means to provide bespoke investment strategies that asset managers can then provide to their clients.

"*Our team consists of 8 people with various backgrounds ranging from computer science to math to financial engineering to business development. We are extremely passionate about what we are doing and in advancing the frontier when it comes to both AI and financial markets*," the startup founder said in an interview.

The founder, who is half-Taiwanese, said that his company has been in Taiwan since 2015 holding a representative of its headquarters in Europe. He recalled that applying to switch from a representative office to a local company was a bit of a struggle.

"*We registered the business by ourselves. It took us approximately one month after we figured out all the steps*," he recalled.

He said that registering a company in Taiwan can be somewhat challenging as a foreigner as it is not as straightforward as in many cases in countries in Europe, where most of it can be done online.

"*In Taiwan, tasks have to be done in a certain sequence and it is not possible to do everything at once. It involves going to several different government offices and handing in documents in person, waiting for each office to complete their tasks in order to take those official government documents to the next government office, and doing the same procedure again. Our greatest challenge was the language barrier, since most/all public information regarding this was in Chinese, and all application forms were in Chinese, and all documents and information had to be provided in Chinese as well,*" he noted.

Tips for registering business by yourself

Just like Be Lucky Taipei, this case proved that anyone can skip the agency route to save cost.

"*If possible, it is highly advisable to have someone that is a local Taiwanese (or knows Chinese very well) to do the company registration and the application for the subsidy. Be prepared to spend time traveling between government offices, that everything is done on paper and in Chinese, and bring the company seal and personal seal for each visit. It is also highly recommended to write the subsidy application in Chinese*" was the founder's advice for those in Taiwan and willing to process the application by themselves.

The mysterious case of subsidy rejection

To be sure, this startup is not new to grants and subsidies, and applying for as many as possible while building a startup is a great way to stay afloat. The team has won among thousands of applicants for the startup incubation program in Korea. They have also won grants from Singapore. On top of these, several financial companies have started testing their product.

The above information has been included in their application for subsidies in Taipei. However, for some reason the internationally recognized startup has failed thrice to get Taipei's subsidy.

"*We applied for the SITI [Subsidies & Incentives for Taipei Industry] 'startup subsidy'* of 1 million TWD," he noted.

"*Writing the application took a very long time as it is very comprehensive. It is also highly advisable to meet the SITI office to have them go through the application before submitting it, as there are technical and bureaucratic things in the application that need to be fulfilled, and if not, then the application will be rejected*," the company founder said.

The company's failed attempts at getting a SITI subsidy were not either technical nor bureaucratic concerns. A letter that they received noted that they failed to provide a market for their product. This when firstly, Taiwan's focus for innovation development has been AI, which is the company's product and secondly, the company has financial institutions already expressing interest in testing the company's product. Still, from what he recalled, there could be a mismatch between the expertise of the judges on the product that he presented, as one reason for failing to get the subsidy. Even before presenting, the founder had sought advice from those who had done it before. There were also people at the same government agency who provided sound advice on how to obtain the grant, he noted.

Expert advice: Fit in the government's box

We sought advice from a company that does consulting with startups, foreign and local. This company's services include marketing, business development, and securing funds. We chose not to name this expert to protect his identity.

In an interview, the expert noted that many startups fail to secure funding from the government subsidies and grants because their strategy is to convince the judges that their startup idea will work according to their own metrics. This, when the judges have already a preconceived idea of what they are looking for before reading a startup's application.

"You have to read through the government's guidelines, and see how your startup can fit those requirements instead of convincing them [the judges] with why you think your company deserves the grant... You have to fit in their box," the expert noted.

The expert admits that there is a gap between what the government says at the policy level and how policies are implemented on the ground. This is because the government, with its strict rules on corruption, works with other government agencies to execute their subsidies and grant policies. For example, the Taipei government has pasted all over its websites that they do not work officially with any private agencies in helping startups secure government funding. This means no guarantee for a success case even if a startup hires a consulting company for its subsidy or grant applications.

UPDATE: Taipei implemented a new rule stating that they would prioritize the applications of those who have already been recognized abroad. Probably, as a result of many whom we know have lodged this similar concern. This company is reapplying again. New rules also state that an application needs to be handed within one year of a company's formation.

Case C: How confusing tax rules could dismay foreigners

TaiwanIt is a resource and a service for Israelis and Jewish travelers heading to Taiwan. Established in 2012, the business has been fully registered as a multi-service business since 2017. Aside from the general tourism and travel services TaiwanIt offers, it also operates to provide Israeli and Jewish businessmen consultation, tours to key locations, and continuing education, particularly in culture and language. TaiwanIt also does the reverse for Taiwanese nationals, particularly companies, that are interested in investing and or traveling to Israel.

TaiwanIt's owner, Meital Margulis Lin, shared experiences in registering the business in Taiwan, which was said to be of average difficulty relative to other places, because they had assistance from a company that specialized in assisting foreign investors in processing the necessary documents and going through bureaucratic procedures to set the business up and acquire the required permits.

"*I feel that it is not difficult to register as a business in Taiwan. Taiwan has special companies that help foreigners to register as a company and take care of the paperwork regarding to this matter,*" Lin who is married to a Taiwanese noted.

Confusing tax rules

Lin however noted that one of the challenges after registering her business was the method taxes were computed in her case, as the company also makes revenue from abroad.

"I feel the main difficulty is all the part related to accounting because in Taiwan there is an accounting method that is a bit complex, especially for companies whose main revenue is from abroad," Lin noted, adding that other foreign business owners relayed to her their confusions on Taiwan's rules on taxes.

When asked for advice for foreign nonprofit firms who wish to open a business in Taiwan, Lin said, *"I know that the regulations in Taiwan are a bit strict and* [we] *need to come up with a big group of members who are part of the organization before it can be registered. I would recommend consulting with one of the companies in Taiwan who are helping non-Taiwanese to open business in Taiwan. In my opinion it saves a lot of time and misunderstandings to have a local company taking care of all the paperwork and the communication with the different authorities."*

Significant progress in Taiwan's ecosystem

Throughout her stay in Taiwan, Lin observed great progress with the startup ecosystem in Taiwan, while pointing out that other things still had to be done.

"I believe there is also still a long way to go. Also, from my conversations with young Taiwanese, I feel that many are afraid to start their own business because of fear of failure and there is pressure from the older generation to find a stable job instead of taking the risk to start something new," Lin noted.

She observes that Taiwan in a way is still very conservative both in education and in mindset, where focus is on respecting rules, elders and hierarchies, as written on by Confucius. "*All of these factors are very important in society but in trying to do something new, there is always a factor of 'breaking the rules' and risk-taking*," Lin noted.

Lin said that she also offers consulting services for local Taiwanese companies that are interested in the Israeli market, and that she holds lectures in universities and private sectors on creative thinking and entrepreneurship in an Israeli style. Lin is also writing a book on Taiwan culture, and her life as an Israeli married to a Taiwanese.

Case D: How a French kid's health wear company got accepted into three Taiwan accelerators

Teams8 is a France-based company with its first health platform dedicated to children. Their first product is a watch that enables kids to do physical activities via superhero games, while being able to check information on their vital health signs.

"*We created the company in France in November 2015. We came to Taiwan mainly to find a hardware partner that will handle the physical parts as our team focuses on software. We also wanted to grow our network here stronger and faster, and to validate the potential we have here, in Taiwan and Asia,*" said Teams8 founder Stéphane Daucourt in an interview.

Teams8 has successfully gotten into Bridges 2 Mass Challenge, WDH Accelerator, Hype SPIN Taiwan.

"*As far I have seen Taiwan is very open to foreign startups and it is willing to support innovation with programs and grants. Big corporations are also willing to receive you easier, even though it is still hard to find the way for them to cooperate with startups* [as in every country]," Daucourt noted.

Because of his experiences, he said they are currently in the process to "*incorporate here, hire a team here, developers and marketing, to work on partnerships, B2B sales and local marketing for B2C.*"

"*I believe that every startup who is handling hardware should have a look at Taiwan as a good option for manufacturers in Asia. I recommend applying to accelerators here as they will know how to give you the local ecosystem and culture,*" Daucourt noted.

How to get in accelerators

When asked for advice on how to successfully get in the accelerators program, he said "*It's* [all about] *personalizing the application to Taiwan needs and goals, among which are to create jobs, generate partnerships between Taiwanese companies and international companies, and finding unicorns.*" For example, reviewing each accelerator's specialty and applying to those where your startup can gain and also add value.

"*The best thing is that most of them don't charge or take equity, even more some offer accommodation and flight trips. I believe Taiwan understands the particular situation for startups and is really trying to find the best way to help, and not just creating an ecosystem to appear big,*" Daucourt said.

He also advised us to join international events hosted in Taiwan. "*There are many events here like InnovEx/Computex, Meet Taipei, etc. that are great to meet partners from all over Asia, you can also apply to program to have a booth, accommodations and flight for free,*" he said.

Chapter 8

Suggested Policy Improvements for Setting Up a Business in Taiwan

Chapter Snippets:

To provide a more comprehensive view of setting up a business in Taiwan, we have summarized some of the key takeaways from a June 2020 whitepaper from the American Chamber of Commers in Taipei. This is followed by interviews with foreigners providing suggestions for what could be improved with the current government policies. The interviews were done for those who are planning or have already gone through the registration of their business. These interviews are separate from the case studies in the previous chapter.

Long-term mindset, short-term deliverables

The American chamber of commerce in Taipei has released a white paper that includes suggestions on how to improve the overall startup ecosystem in Taiwan. The paper suggested building an environment that favors competition, attracts investment, and nurtures talent.

The paper said that if Taiwan wants to catch up with countries like Israel, Estonia, and Singapore, it needs to set ambitious total investment targets in startups by increasing public and private investment in startups. One of the ways to sustain economic growth is by setting an annual target for startup investments based on a given percentage of GDP.

"We urge the Taiwan government to embrace a long-term mindset with short-term deliverables, bearing in mind that an innovative environment relies on competition, investment, and talent," the paper said.

The paper also suggested encouraging the free circulation of startup talent. "An innovation-based economy depends fundamentally on having a robust, internationally connected startup ecosystem. It is therefore vital to remove restrictions and create incentives to encourage the free circulation of startup talent. Startups with international connections and international team members are more likely to succeed in global markets," the paper noted.

While reciprocal arrangements with other regional partners would be ideal, such as the proposed creation of a "G-Asia pass" aimed at extending national treatment to foreign startup talent, it suggested that Taiwan should offer such treatment to all foreign startup talent willing to base some or all of their activities in Taiwan.

"Further, Taiwan should extend the Gold Card program to foreign startup talent and devise other incentives to attract such talent. These incentives could include tax reductions (perhaps in proportion to the number of local jobs created), a flexible regulatory environment through sandboxes to enable disruptive innovation, and reduced educational costs for their children," the whitepaper said.

Interview 1 - Tomás Swinburn, Ireland

Tomas, a digital marketer, recruiter, writer and community coordinator, came to Taiwan originally as a scholar student at National Chengchi University (NCCU) in Muzha and graduated with a Master's Degree in International Communications in summer of 2018. MillionDC sent him questions regarding his interests in starting a business here in Taiwan.

Q: Are you planning to set up your own company here?

A: I'd like to set up a small company in Taiwan, but without an open work permit, I simply do not have the scale of investment needed.

Q: What are the barriers that hinder foreigners from smoothly setting up a business in Taiwan? It could be based on your experience or as relayed to you by other foreigners. For those told to you, was this in a forum or a foreigners' group event?

A: Pretty much, everything from government requirements, banks, bureaucracy, and the lack of interest from both a government and business level is hindering many expats from opening businesses in Taiwan. A big reason for many of my friends to not consider Taiwan is because to keep an entrepreneur visa you need to either spend a certain amount of capital in a year or make a certain amount of profit. So, for a startup in its first year, it either needs to make a profit or put itself in debt to keep the visa for the founder. It's insanity. What startup do you know has made a profit in its first year?

I had two job offers that were taken away because the companies couldn't meet the minimum capital requirements to hire a foreign national.

These were companies that wanted to internationalize and they couldn't because they couldn't keep a certain amount of money in the bank. Basically, the government is trying to say "to hire one foreigner, you need to restrict your access to capital in a bank" and pay us more. It doesn't give companies an incentive to hire international talent unless they're large enough that the capital requirement doesn't matter to them. But, startups in Taiwan are the most foreign-friendl places to work. Large companies here aren't friendly to locals, why would they be friendly to foreigners?

Q: Could you recommend some policies that the Taiwanese government should implement to make the process of registering smooth?

A: I'd say the Gold Card scheme needs to have more realistic requirements. For example, you can get a gold card simply by making over NTD 160,000 in a month. That means that someone from high-income countries with less experience can get gold cards when compared to lower-income country professionals that have better experience. This is just one example. Also, for the gold card, many foreign nationals are submitting phantom income to make their income look higher than it was, so the system is already pretty easy to cheat. No, the government is copying a Singaporean model that simply does not apply to Taiwan because, as mean as this sounds, Taiwan is not Singapore. It doesn't have the same job opportunities, salaries, or industries. Taiwan needs policies that are realistic.

Q: Could you cite other policies that would attract foreigners to set up a business in Taiwan?

A: Current policies are working, but they need to lower thresholds and not discriminate against people from lower-income countries. Likewise, the government needs to open up Taiwanese citizenship to foreign nationals without needing them to relinquish their previous citizenship. What's the point of setting up a business in Taiwan if you can't vote, get a bank loan, get a phone plan without a deposit? It's not a good look for Taiwan that foreign nationals, business owners or not, still have to get a Taiwanese friend to help, even if they create a successful business. Again, many foreigners come to Taiwan and set up businesses, but to keep them, an APRC isn't enough.

Interview 2 - Franck Delecroix, France

The overall safe and progressive environment for foreigners had attracted Franck Delecroix of France, to set up his business in Taiwan but he had a change of heart out of frustration that he hopes gets resolved by the government. Delecroix responded to an online poll by MillionDC in full details.

“I wanted to open an eCommerce business for French products and use Taiwan as a platform for Asia. I was attracted to Taiwan by this stability, middle cost to stock product, emplacement and skilled people. And I was ready to invest more as the minimum capital requirement, and also hire Taiwanese. After 5 months of making a business plan and meeting a lot of potential partners in Taiwan, I decided to open my company in France and used HK as a platform for my product,” Delecroix said.

Requirement for keeping a Taiwan visa

“The requirement after the first year of NTD 3 million is one of the main issues. Not every kind of business can achieve it... the first year [of operation] is focused on finding customers. I decided to not invest in Taiwan, because it was too much risk for me to invest money and time,” Delecroix noted.

Other foreigners polled by MillionDC shared the same sentiment. “NTD 3 million to reach is too high for smaller businesses,” said Guillaume Douce, a French owner of a business in Taiwan.

Another point of view raised by other foreigners in the same poll was that the set minimum amount was based on a fixed notion of the type of businesses that Taiwan is supposed to attract in order to spur the country's economic growth.

Banking concerns

Delecroix also noted that Taiwan is not sufficiently connected to Asia. "Taiwan is free, not too expensive, stable, has educated and skilled people, and a very strategic place. It could be the perfect platform for pan-Asia strategy. Moreover, when HK is unstable and very expensive," he said.

"Sadly, exchange rates in the bank are so expensive, Taiwanese potential only proposes service for Taiwan, not Asia, Internet payment gateway for Taiwanese companies only accept NTD and USD, delivery for B2C in Taiwan to other countries are almost nonexistent or very expensive," he noted.

Other foreigners polled by MillionDC had the same concerns. "Banking is a huge obstacle here. Too many procedures, outdated old systems and too slow," said Michael M. Thomsen, a Dane living in Taipei. Justin Van Midden from Australia suggested that banks not "discriminate against foreigners... For example, allowing them access to credit in the same way Taiwanese people can."

Long transactions in banking is another concern for Sam Crawford, a British owner of branding company Iced Copy.

"With banks it can take a long time to do very simple things, at least at mine anyway. More than once I've waited over 45 minutes to do an international transfer, and that's not including the time waiting in line beforehand.

Lots and lots of paperwork, checks and forms. Also the same case with receiving transfers. I received a SWIFT transfer from a client but they didn't put LTD at the end of my company name so my bank sent the money back to the client and we incurred significant fees for them doing it," noted Crawford.

Unclear laws

Delecroix also highlighted the need to be clear with implementation of the laws related to setting up a company. "Laws are not clear, too much grey area," he said, adding that new foreign companies doing the same businesses with those created in the past and still running, would be considered illegal.

"The minimum requirements to partner with Taiwanese company are often very high, making it difficult and risky for a new company to start," he noted.

The *guanxi* problem

We've discussed the importance of *guanxi*, and Delecroix highlighted one key element about this key Taiwanese cultural trait. "Guanxi, guanxi, guanxi. If you do not have guanxi, things are going to get so slow and complicated. It takes decades to build a real *guanxi* network," he noted.

Interview 3 – Sam Crawford, UK

Sam Crawford, a British owner of branding company Iced Copy, highlighted a need for a well thought out, comprehensive guide to setting up a business on the government website. "Currently, it's poorly laid out, hard to find, hard to decipher and seems to exist across a range of different sites," he said in a poll conducted by MillionDC.

Long timeline to set up a business

Sam also said reducing the time it takes to set up a business would be a big improvement. "For me, it was 4 months, and that included a lot of running around between different departments," he noted.

He also suggested having properly sign-posted stipulations. He noted that the regulations often seem to be "inexplicably draconian for very minor points, and unfathomably lax for other major ones" and that being strict and lax seemed "depending on which staff you're dealing with that day."

Better communication within government departments

Sam noted that there seems to be poor communication within government offices, both across different departments and in the same ones.

"For example, six weeks after handing my application form in and not seeing progress, I enquired in person at the MOEAIC what the holdup was. They said they hadn't received my form from the other department yet, so I headed there to be told they sent it weeks ago.

I headed back to MOEAIC and after explaining for 10 minutes that one of the departments had to be wrong, they checked again and found my form in an archive folder so it hadn't been looked at. That's one of numerous hurdles which couldn't be foreseen despite the planning I did," he recounted.

Sam also shared the same sentiment of many foreigners with the business visa requirement of turning in NTD 3 million in revenue in the first year.

"Currently they're very tough. One year to meet three million [NTD] revenue when it can take 4 months to set up the company from the moment you have your initial visa, and that's if you already have an accountant sorted. In the case of my visa I had to apply 45 days before it expired (ended up taking 65 days for them to renew it), so that leaves 6-7 months to turnover three million NT in a new country with a new business, unless you get three full-time employees or manage to jump through hoops another way (which I did, and it's a lot of hoops)," he noted.

He said better banking systems would be good too. "With banks it can take a long time to do very simple things, at least at mine anyway. More than once I've waited over 45 minutes to do an international transfer, and that's not including the time waiting in line beforehand. Lots and lots of paperwork, checks and forms. Also the same case with receiving transfers. I received a SWIFT transfer from a client but they didn't put LTD at the end of my company name so my bank sent the money back to the client and we incurred significant fees for them doing it." He said.

In the same poll, Paul Shantz from Canada noted that banking issues could be more political and cultural.

"It's not that hard, they just need to cut the bureaucracy in the banking system... Singapore and Hong Kong have done this, decades ago... the problem is deep down the people here are protectionists and don't want a huge influx of foreign capital. So, it's more of a political and cultural problem than a 'how can we do it?' problem," he explained.

He also said, "Anyone who wants to do the same I recommend paying someone experienced and knowledgeable to do as much of it as they can for you, because otherwise it eats up so much of your time. I didn't do that - my bad."

Chapter 9

Key Concepts in Building a Startup

Chapter Snippets:

As another guideline in setting up your business in Taiwan, this chapter highlights some lessons along the way from conceptualization, finding a co-founder, building a minimum viable, product to pitching to investors. We begin by introducing who we are and our vision, followed by the lessons that could be valuable. This chapter is culled from another book that I wrote titled Cyberpreneur Philippines. To begin with, MillionDC is a startup that is based in Taiwan with aim of helping startup founders become successful with their goals.

About MillionDC

Million Dollar Concepts (MillionDC.com) is a learning platform for entrepreneurs in the developing countries, based in Taiwan. The requirements of entrepreneurs from developing countries are way different compared to developed countries, they need education more than funding. This is why we offer free workbooks, and provide workshops on how to turn ideas into profitable concepts. We believe in the saying "Give a man a fish and you feed him for a day. Teach a man to fish and you feed him for a lifetime".

Based on our experience here in Taiwan, this chapter will give you some basic tips on how to conceptualize your startup, making it a reality, and pushing it to pitch competitions. These tips seem to be easy to execute but most founders tend to ignore them as the pointers seem counterintuitive while others appear like they don't directly contribute to improving your product.

Being awarded as Top 100 Startups in Asia by e27 in Singapore is an important validation of what our team at MillionDC.com worked hard for. The road to this milestone is not easy - we almost gave up amid criticisms, rejections, and my depleted savings. Looking back, we realized that this opportunity was not handed out of luck but a result of the small things that I have been practicing from day one. The key steps are both done offline and online.

Lesson 1 - Never stop talking about your idea whenever there is a chance.

Contrary to what some people think that talking about your idea will get people copying it, we openly discussed my startup from day one. By discussing with people from different backgrounds, we got well-rounded feedback, which helped me solidify my concept. We used the feedback to develop our website. Talking about my concept also served as a practice for an elevator pitch. It helped me convey my ideas in the simplest form, using words that anyone who speaks English at basic level will understand. Discuss it with your friends, and let them critique your idea. You will not hear good words for sure, as your friends are supposed to be frank if their goal is to help you. Take it all in as constructive criticism. When something is not clear during, try to restate what you heard and ask them if you got their point right. When you are confident enough, bring your idea to people outside your network and do the same. Over time, you'll discover that your idea becomes a concept. To be sure, a concept is usually made up of 1. consumer insight or an accepted consumer belief, 2. Reason to believe - specific ingredient or function, and 3. Benefit statement. While an idea is simply an unrestrained thought about a certain topic. MillionDC.com platform as we envision, will help entrepreneurs turn their ideas into workable concepts.

Lesson 2 - Limit your networking

We've attended only three startup events while founding MillionDC. The first time was to test our hypotheses that there is a need for an online social platform and that people are willing to share their thoughts about their idea on this platform.

The second one is to do an impromptu pitch before a crowd, even when we don't have an MVP yet. On the third time, we realized that these types of events simply dragged down our efforts, more time networking meant less time working with the team about improving the product. Limiting networking allowed us to focus on work. We told ourselves, we'd come back when we have something new to say such as the availability of an MVP. Choose the networking that you will go to and you should have a reason for showing up to an event, other than listening to a pitch. If your goal is to listen to pitch, you can simply google that. One good reason is to get a chance to talk to VCs, they are sure going to help you turn your idea into an actionable concept. After talking to some, stop networking for a while and work on polishing your idea, or build your MVP.

Lesson 3 - Don't aim for perfect MVP, go cheap

Since our startup back then was a social media platform, our first minimum viable product is simply Facebook Fan Page (www.facebook.com/milliondcltd). In it, we tried experimenting with what kind of content our fans, or potential users would like to read. We studied the behaviour of our target fans by regularly posting content, and changing the content to see how fans would react. The Facebook site gave us some validation, that people need a dedicated page for content about starting a business, and the how-tos. We then used this information in building the beta version of our website, www.MillionDC. By the time we applied for pitch competition, our fan page had about 1,400 fans and was still growing. Our website however was not even gaining traffic and some of the features were not even functioning.

And so, we were surprised that the review committee at the Top100 Qualifiers was actually looking at our Facebook Page as among the considerations for the shortlist. We also utilized other free available platforms such as Google Plus, Twitter, and LinkedIn. As much as possible, use the free resources that are available when building your MVP. Remember that the faster you can build it, the faster you can test your hypothesis. In the book Lean Startup, Eric Ries pointed out the importance of minimizing total time through using Build-Measure-Learn Feedback Loop. We were able to do this by going for a cheap and easy to execute MVP, instead of coming up with something expensive.

Lesson 4 - Choose your co-founder wisely

A co-founder is your co-pilot for running your startup. He should have an area of expertise that you don't have and that's one key reason you need a co-founder for. Don't go for pure talent and with an attitude. Get someone who has the same passion as you have for your business. When your potential co-founder asks about how you plan to split the ownership of your future company, drop him at the very instance. Some of my friends asked for co-founding MillionDC.com but had politely rejected the offer to keep our friendship. Remember that people are not the same as friends, as co-workers and more so as business partners.

Eventually, you and your co-founder are going to talk about ownership of the business. Here is a good tool we found at http://foundrs.com/ for your reference.

Lesson 5 - When pitching, tell the story of your potential customers.

You will only have one chance to pitch. Convincing your potential investors is about telling a story of how your potential customers will use your product. During our pitch at the Qualifiers of Asia's Top 100, we didn't bother the judges (all of them were VCs) with numbers. Instead we walked everyone with the process of using the site and why they need it. It became easy for the judges to understand our product, and as a result, the judges were able to give us much valuable feedback. Remember that the judges themselves are possible users of the product. They should be able to grasp your concept well, and from that they would be able to identify it is viable and worth investing. Days after our pitch, we got some investors asking about our product and had arranged separate meeting with us. Know your product well, that's the key for pitching. You should be able to discuss the product, its benefit to users, and its value to investors. You should be able to separate your product from existing similar products. There should be some other similar products, try googling it with your product's basic keyword. VCs would suspect that you don't know your product well if you can't even identify a potential competitor. They'd think that you did not do your homework well.

Here are the criteria set by the judges at e27's Top 100 Startups:

Criteria	Max Points	Guidelines
Problem and Solution	5	Product design, technical finesse, and creativity of the solution, Problem solution fit.
Market	5	Size of market, Market opportunity and fit
Revenue Model	5	Creativity, feasibility, scalability of revenue model
Team	3	Ability to execute, potential, past success, the X-factor.
Pitch	3	Clarity of presentation, creativity of presentation, ability to answer questions effectively

Chapter 10

How to think globally for your startup

Chapter Snippets:

While your startup is established in Taiwan, it is already exposed worldwide and it has to be competitive globally, to survive. This chapter provides some concepts to make sure that you are globally for your business. This chapter enumerates the lessons based on building MillionDC, from ensuring high quality products, identifying competitors, to marketing your product worldwide via the digital space. This chapter is culled from another book that I wrote titled Cyberpreneur Philippines.

Even if you're based in Taiwan, you are already throwing yourself into the global competition from the very moment that you launch your online business. You have no choice but to take part into the wide-scale battle as it could be seen by the whole world wide web. In this digital era, your idea is never unique and the rule of thumb is that for every idea that you have, three people in another part of the globe are already working on it and with a more solid concept. However, this era also gives any business an equal playing field to compete internationally with vast available information that can be accessed for free.

Ever since MillionDC.com was conceptualized, we have always thought of it as a global brand and not a simple mom and pop shop. This framework of thinking allowed us to acquire ideas and innovate to bring my game to the global level.

The following tips will allow you to prepare and bring your startup at an international level.

Lesson 6 - Check your product first and foremost

An honest evaluation of your product is the very first approach to preparing your business for an international competition. As we have often said based on readings, there is no unique idea and that your business model has something similar with another business somewhere else. Google these businesses and study them well. As part of your business plan discussed in Chapter 4, you will need to have a SWOT analysis. This should be done both for your business as well as for your competitors. This tool allows you to see your business' key strength as well as weakness against competitors.

Upon doing this analysis, you will realize that each of your competitors has its unique selling proposition (USP) and you should have one too. Your USP will be your main selling point when you start marketing your brand globally. USP's don't have to be feature centric. Take the case of Toms shoes. The approach of giving to the less privileged for every item sold is its USP.

The SWOT is also a good tool to trim down your business. Yes, you read it right. At the very early stage of your business, you will have to trim your product line to ensure that the quality can compete globally. List down the product offerings of your company and your competitors. For a website, you don't need too many tabs when you cannot fill them all with valuable content. For instance, at MillionDC.com we realized that including a photo editing tool for our bloggers was not necessary especially that our plug-in was buggy at the moment. We also took out a lot of sections and kept the site neat and easy on the eyes. Remember, you are catering to the world as your audience and so a simple and understandable product is better.

You will need to have someone outside your circle to evaluate your product to avoid bias when doing SWOT analysis, coming up with your USP and trimming down your product line.

Lesson 7 - Strategize before jumping into the ring

Having done your SWOT analysis for your company and competitors will help you formulate your strategy. Ask which markets do your competitors have a strong presence?

For someone who is just starting, avoid those markets and find a niche, a place or group of people that your competitors are not focusing on because those areas are somehow negligible for their scale. Avoid getting into your competitors radar by tricking them into thinking that either you are too small or that you are not a direct competitor. Slowly creep into their market until you are confident enough to fight head-on. Many times, great ideas from startups are being killed by competition because big companies can easily replicate these ideas. Having a clear and actionable strategy is very important before jumping into the battle.

Lesson 8 - Co-work with other startups

Co working with other startups will help you think globally. Being in a group of Zuckerberg-wannabes allows you to find potential partners for entering a different territory or complement each other's business model. Look for common points of interest with another company and see how you can both benefit either from a joint marketing campaign. Try getting a spot on co-working spaces, rent is cheap for sure. Being in co-working space can provide great ideas and could also lift your startups credibility.

Lesson 9 - Think big, fake it till you make it

Part of taking your business on a global scale is getting investors. Remember that you are the first person who should believe in your business and if you are not confident in your success, no one else will be. To entice investors, you will need strong convincing power and sometimes the only way to convince them is to say that your plans are already executed. This also helps you to push people onboard and make reality catch up with your story.Don't ever think that you are a small business because by doing so you will belittle your growth and not much will be achieved over time. Think big by having a clear vision of how big your company will be and coming up with a map on how to get into that vision. This means reading a lot about the industry projections and how you could leverage on growing trends. This is a good story to tell investors, it means you did your assignment.

Faking doesn't mean lying or bragging. According to TechniAsia, common lies include faking a valuation, claiming other startups' success as your own, and fudging one's usage metrics with bots. At the Echelon Summit Asia in Singapore, we were asked by VCs about my website's traffic and we honestly told them that it's not for me to show because it's low and that's because it was relatively young (two weeks at that time). We could have used bots to pump up the numbers before attending the event but we know that these VCs are a lot smarter than that and that they appreciate honesty and being confident.

Lesson 10 - Get your social media presence globally-oriented

Social media is the most useful free marketing tool that startups can leverage on. But most of the time this is not a priority of founders and that they use it only for creating hype. The power of word-of-mouth marketing through social media has made it possible for small startup businesses to reach millions of consumers with the click of a button. A globally oriented social media presence is one that has content that can cross borders in terms of culture, religion and belief. Choose the appropriate social media platform for your business. Being on only one but well-managed platform is way better than being on many sites that are not managed properly.

If you have a post that is for a specific country, you have to use the targeting tool on your social media platform to avoid people from other countries unliking your page. Make sure that your content is not hard-sell, imagine the lifestyle of your customers and create content around it. For example, at MillionDC.com, we have posts that answers basic questions for those who want to start their own business. Sometimes we post inspiring quotes about launching a startup, post things that are shareable.

Chapter 11

Miscellaneous Steps

Chapter Snippets:

This chapter looks into further activities that foreigners would be doing after registering their business and so this is placed at the very end section of the book, separate from the steps in establishing an enterprise. It covers the guidelines in obtaining construction permits based on a case for building a warehouse in a World Bank document titled "Doing Business 2020" and Tax Payments based on information from the Ministry of Economic Affairs.

Obtaining Construction Permits

Step 1 - Obtain information about infrastructure from the water company

Head to Taiwan Water Corporation at One-Stop Counter to obtain information on the intended building site's water and electricity infrastructure. It would take you 3 days to process some papers to complete the agency's requirements.

Step 2 - Obtain information about electricity infrastructure from electricity provider

Head to Taiwan Power Company (Taipower) at One-Stop Counter to obtain information about electricity infrastructure. This is not a mandatory step and takes 3 days to process.

Step 3 - Request and obtain building permit from the City Government

Head to the City Government One-Stop Counter for required city government approvals.

"Before presenting an application, the applicant should carefully follow the Self Checklist of Application for Construction Permit and Review of Design for Water Supply Documents (OSC2)." Doing Business 2020...Once all amendments have been made, a construction permit is issued, and fees for the construction permit, specification of building setback line, wastewater discharge permit, and land ownership certificates are paid," Doing Business 2020, World Bank.

The World Bank document noted that it is possible to apply for a building permit online via http://tccmoapply.dba.tcg.gov.tw:8080/tccmoapply/. It would take 21 days to complete this step and fee of about NTD 31,000.

Step 4 - Report the start date and present construction plan to the City Government

At the City Government One-Stop Counter, report your construction start date and pay air pollution protection fee before starting construction. Ahead of time, prepare the application for commencement of construction, a construction plan, and other required documentation. This step would take about one day and fee of around NTD 25,000.

Step 5 - Request occupancy permit, post-construction approvals and registrations

After the construction is completed without involving damage to adjacent property, take the original construction permit and make a one-time submission of the Self Checklist for Post-Completion Application Documents (OSC4) and apply for an occupancy permit at the City Government One-Stop Counter. The mode of submission is the same as the construction permit application. If the building is found to be following applicable regulations, the One-Stop Counter will approve the issuance of an occupancy permit and will require the applicable fees. This step would take a day and about NTD 79,000 fees.

Step 6 - Receive final inspection

As part of the review process, the relevant departments will conduct a joint final inspection which could take place in one working day and free of charge.

Step 7 - Obtain occupancy permit, post-construction approvals and registrations

Once the issuance of the occupancy permit is approved, the One-Stop Counter will transmit the

documentation to apply for registration of title to the local land administration office. It would take 33 days to get this processed and there are no fees required.

Step 8 - Apply for water supply and sewerage connection from Taiwan Water Corporation

The applicant may channel the request for water connection through the one stop counter for building permits or may request it directly from the Taiwan Water Corporation. This step would take one day without any required fees to settle.

Step 9 - Receive water and sewerage inspections

The Taiwan Water Corporation will conduct an inspection to evaluate that all conditions have been met and match the specifications submitted in the Review of Design for Water Supply Documents (OSC2), in one day without any fees to be paid.

Step 10 - Obtain connection to water and sewerage

After the water and sewerage inspections are done, the warehouse will receive water and sewerage connections. This step would take about 19 days and fees of about NTD 25,000.

Paying Taxes

Based on the Ministry of Economic Affairs website, a profit-seeking enterprise is subject to profit-seeking-enterprise income tax. By its definition, profit-seeking enterprise is an entity established in the form of a sole proprietorship, partnership, company or other form of organization that operates for profit-seeking purposes through a fixed place of business, regardless of whether the enterprise is owned by the government, private sector, or jointly by the government and private sector.

Profit-seeking enterprise tax (sometimes called corporate income tax in other countries), follow the following rates:

- For taxable incomes NTD 120,000 or less, no income tax liability is incurred.

- For taxable incomes above NTD 120,000 the tax payable is 20% of total taxable income to a maximum not exceeding 50% of the portion of taxable income in excess of NTD 120 000.

It noted that profit-seeking enterprises with a taxable income of not more than NTD 500,000 are subject to annual adjustments with a tax rate of 18% for the year 2018, a tax rate of 19% for the year 2019 and a tax rate of 20% for the year 2020 and after, but income tax liability may not exceed 50% of the portion of taxable income over NTD 120,000.

The government also listed a number of exemptions for paying tax. Among others, those that are specifically for foreign enterprises are:

i. Business income derived from the operation in Taiwan of a foreign enterprise engaging in international transportation, provided reciprocal treatment is granted by the counterparty foreign country to a Taiwan international transport enterprise operating in its territory.

ii. Royalty paid to a foreign enterprise for the use of its patent rights, trademarks, and/or various kinds of special licensed rights in order to introduce new production technology or products, improve product quality, or reduce production cost under the approval of the competent authority as a special case.

iii. Remuneration paid to a foreign enterprise for technical services rendered for the construction of a production facility for an important manufacturing enterprise as determined and approved by the competent authority.

iv. Interest derived from loans granted to the Taiwan government or legal entities in Taiwan by a foreign government or international financial institution for economic development, and interest earned by foreign financial institutions from the financing of resources offered to their branch offices or other financial institutions in Taiwan.

Compliance Requirements

Annual income tax returns are due during the period 1 May until 31 May, for a company with an income tax year aligning with the calendar year. For companies with income tax years that do not align with the calendar year, filing is due 5 months after the year-end.

Businesses are required to use NTD as a bookkeeping base, it is required to convert and prepare the tax return in NTD if the bookkeeping base is in foreign currency. cxcxcxcxcx

Sources

2019 Index of Economic Freedom, The Heritage Foundation – https://www.heritage.org/index/pdf/2019/countries/taiwan.pdf

2019 Taipei International Design Award "Design Taipei; Dream Taipei" – Key Design Concepts - https://www.taipeiecon.taipei/english_cont.aspx?MSid=1037362756061306441&MmmID=4002&CatID=744725544402432171

2020 Taiwan White Paper, Taiwan Business TOPICS, American Chamber of Commerce, Taipei – https://amcham.com.tw/advocacy/white-paper/

Abso Capital – http://www.absocap.com/

Act for the Recruitment and Employment of Foreign Professionals, NDC – https://foreigntalentact.ndc.gov.tw/en/

Action Plan for Enhancing Taiwan's Startup Ecosystem, NDC – https://www.ndc.gov.tw/en/Content_List.aspx?n=DD6EB3B5F084F394#:~:text=The%20Action%20Plan%20comprises%20five,Taiwan%20an%20Asia's%20startup%20nation.

Action Plan for Welcoming Overseas Taiwanese Businesses to Return to Invest in Taiwan, NDC – https://www.ndc.gov.tw/en/Content_List.aspx?n=286FD0E985C0EA44&upn=C8BDB84847E24D6B

BeLucky Taipei – www.beluckytaipei.com

Bilingual Nation, NDC – https://www.ndc.gov.tw/en/Content_List.aspx?n=D933E5569A87A91C&upn=9633B537E92778BB

Business Startup Award – https://startupaward.sme.gov.tw/Home/Index/#award/

Do's and Don'ts, Taipei Economic and Cultural Office in New York – https://www.roc-taiwan.org/usnyc_en/post/132.html

Doing Business in Taiwan, Grant Thornton, Taiwan – https://www.grantthornton.tw/globalassets/1.-member-firms/taiwan/media/tw_images/publication-pdf/miscellaneous/2018-01.pdf

Expat Insider 2019, InterNations – shttps://cms-internationsgmbh.netdna-ssl.com/cdn/file/cms-media/public/2019-09/Expat-Insider-2019_The-InterNations-Survey_0.pdf

Fastart – https://invest.fastart.com.tw/en/foreigner-set-up-company-in-taiwan-2/

FDI China, Company Registration in Taiwan –- https://www.fdichina.com/company-setup-in-taiwan/

Hongwell Group – http://www.hongwellgroup.com/#

https://www.moeasmea.gov.tw/article-tw-2403-4074

I rode the subway in Taiwan and saw why it's one of the best mass transit systems in the world, Insider – https://www.insider.com/taipei-metro-vs-america-2019-1

Iced Copy – https://icedcopy.com/

IMD World Competitiveness Ranking 2020 – https://www.imd.org/news/updates/IMD-2020-World-Competitiveness-Ranking-revealed/

Invest Taipei Office – https://invest.taipei/pages/E_main.html#

IoT Integrated Service Center (IisC) – https://iisc.org.tw/en/index.php

ITRI – https://www.itri.org.tw/english

MillionDC – www.milliondc.com

Ministry of Education Universities Graduates Entrepreneur Program (U-START) – https://www.yda.gov.tw/Content/QandA/contents.aspx?&SiteID=563655426603362361&MmmID=746301557642302131&SSize=10&MSID=2017120410270897234

National Development Fund Startup Angel Project – https://www.angelinvestment.org.tw/

Profit-Seeking Enterprise Income Tax – https://investtaiwan.nat.gov.tw/showPage?lang=eng&search=55

Service Industry Innovation Research (SIIR) – http://gcis.nat.gov.tw/neo-s/Web/default.aspx

Small Enterprise Loan – https://www.moeasmea.gov.tw/article-tw-2403-4076

SME Innovation Development Project Loan

t.Hub Taipei – https://www.t-hubtaipei.com/

Taipei City Industry Incentive Subsidy Project – https://www.industry-incentive.taipei/page-about-en.aspx

Taipei ranks 6th in PICSA inclusive prosperity index, Taiwan Today – https://taiwantoday.tw/news.php?unit=2,6,10,15,18&post=166428

Taiwan SME Innovation Award – https://www.moeasmea.gov.tw/article-en-2612-4471

Taiwan: Business Practices, Santander – https://santandertrade.com/en/portal/establish-overseas/taiwan/business-practices

TaiwanIt – www.taiwanit.net

Taiwan's advances in digital health care have contributed to global disease prevention and universal health coverage – https://www.biospectrumasia.com/news/55/15136/taiwan-emerges-as-pioneer-in-strengthening-global-healthcare.html

Teams8 – https://team8.tv/

The Asia Silicon Valley Development Plan, NDC – https://www.ndc.gov.tw/en/Content_List.aspx?n=90BEB862317E93FC&upn=7B70255F66FB9DF5

Vice president opens base for 'Asian Silicon Valley', Taipei Times – http://www.taipeitimes.com/News/taiwan/archives/2016/12/27/2003662002

World Bank. 2020. Doing Business 2020 – http://documents.worldbank.org/curated/en/688761571934946384/pdf/Doing-Business-2020-Comparing-Business-Regulation-in-190-Economies.pdf

Worldwide Tax Summaries, Taiwan, PwC – https://taxsummaries.pwc.com/taiwan/corporate/tax-credits-and-incentives

About MillionDC Ltd.

Founded in 2013, MillionDC.com is a learning platform for entrepreneurs. The Taiwan-based company aims to provide educational materials, conduct workshops, and provide relevant resources for startup founders in developing nations with limited resources. In 2015, the company was awarded as one of the Top 100 startups in Asia by e27 in Singapore, followed by another plaudit in 2016 as one of the Top 15 startups to watch by the same agency. The insights for building MillionDC.com have been featured in a book titled Cyberpreneur Philippines, which is co-authored by Mr. Lising and a Philippine National Book Awards finalist in 2016. MillionDC Ltd. (柏創行銷有限公司籌備處) is officially registered in Taipei, Taiwan in 2020.

樂繽紛 46

Startup Taiwan

作　　　者／Paolo Joseph L. Lising
總　編　輯／何南輝
責 任 編 輯／李承軒
行 銷 企 劃／黃文秀
封 面 設 計／Benjamin Chaumeny
內 頁 設 計／上承文化

出　　　版／樂果文化事業有限公司
讀者服務專線／（02）2795-3656
劃 撥 帳 號／50118837 號　樂果文化事業有限公司
印　刷　廠／卡樂彩色製版印刷有限公司
總　經　銷／紅螞蟻圖書有限公司
地　　　址／台北市內湖區舊宗路二段 121 巷 19 號（紅螞蟻資訊大樓）
電話：（02）2795-3656
傳真：（02）2795-4100

2020 年 9 月第一版　定價／650 元　ISBN 978-957-9036-28-3
※ 本書如有缺頁、破損、裝訂錯誤，請寄回本公司調換